Contents

Chapter 1

Best Brains

Best Brains, Inc. (first known as *Hair Brain Productions*) is an entertainment company based in Eden Prairie, Minnesota. It is best known for the creation and production of the comedy/sci-fi television program *Mystery Science Theater 3000*, aka *MST3K* (1988–1999). The company ceased producing the program when it was cancelled by The Sci-Fi Channel in 1999 and closed its studio. The company then functioned as the holder of the MST3K brand for negotiating home video releases of the show on Shout! Factory and its availability on streaming services like Hulu.

According to Joel Hodgson, the name "Best Brains" came from "...a phrase I found in a magic catalog. It was the old 'Vick Lawston' magic catalog. The copy read something like: 'from some of the Best Brains in the magic business!'"[1]

On November 5, 2007, Jim Mallon and Paul Chaplin of Best Brains revived the *MST3K* brand by launching a new series of Flash-based animated shorts featuring the robots of *Mystery Science Theater 3000* on the official website. The series was ended after several months due to cost issues.

On November 10, 2015, Shout! Factory announced it had purchased MST3K and its associated intellectual property from Best Brains for an undisclosed sum.[2] On the same day, MST3K creator Joel Hodgson announced a crowdsourcing campaign to revive the program.[3]

1.1 External links

- Official Site for Mystery Science Theater 3000 and Best Brains, Inc.

1.2 References

[1] "Enlightening Email from Joel". 2011-05-02. Retrieved 2011-05-14.

[2] http://deadline.com/2015/11/mystery-science-theater-shout-factory-1201617524

[3] https://www.kickstarter.com/projects/mst3k/bringbackmst3k

Chapter 2

Cambot

Cambot is one of the fictional robot characters on the *Mystery Science Theater 3000* television series. It is through Cambot's "eye" that viewers watch Joel Robinson (later Mike Nelson) and the other robots as they watch the movies that are sent to the Satellite of Love each week.

2.1 Appearance

Cambot is only seen during the "Robot Roll Call" portion of the opening credits, often with his name reversed, presumably to imply he is shooting his own image in a mirror (this was actually proven in the Season 2-5 intro, when Joel is seen tapping on a mirror that Cambot is in). His appearance was changed with almost every reshooting of the opening credits.

In the original KTMA season, Cambot was depicted as a robot operating a separate camera. Season 1 Cambot was a modification of the KTMA Gypsy, with an integrated camera, whereas later Cambot incarnations more closely resembled television recording equipment. In episode #507, *I Accuse My Parents*, Gypsy presented a drawing that depicted the Satellite of Love's crew as her "ideal family"; in the drawing, Cambot's torso was shown as a long and snake-like tube, not unlike Gypsy's.

Midway through the fifth season of the series the opening was once again reshot, and Cambot was again redesigned, this time with a more compact shape, becoming a round hovering ball with a TV camera vidicon sensor for an eye. He would keep this form for the remainder of the series, although the color scheme was changed during MST3K's switch from Comedy Central to the Sci Fi Channel (becoming blue instead of gray).

2.2 Overview

Cambot acts as an audio-visual conduit between the crew of the Satellite of Love and their observers. He also ap-

parently joins Joel, Mike, Crow T. Robot, and Tom Servo in the theatre when a movie is shown, and records the cast watching the film. Apparently the footage he shoots in the theater is what the Mads end up seeing, as evidenced by Frank once saying "You think you have it bad? We have to watch *you* watching the movie." Although a number of episodes depict the cast reacting as if traumatized by a particularly bad movie, Cambot suffered a severe reaction only once, weeping when several security cameras were systematically destroyed by the hero in episode #620: Danger!! Death Ray. (This was signified by a watery effect over the screen image.) Another rare case of Cambot interacting during a movie segment came in episode #202: The Side-hackers, when Cambot added a mock ESPN scorecard on one side of the screen during the movie's race scenes.

Cambot also frequently provides music, video clips, and other enhancements to host segments. When Joel or another character requests to see through "Rocket Number Nine" (the ship-mounted camera that allows the crew to see the ship's exterior and anything in its vicinity), it is Cambot who provides the image. During the first five seasons, when Joel or Mike would read fan mail sent to the show, they would request Cambot to put the letter on "still store," freeze framing on a close-up of the letter.

Cambot was voiced a single time during the original KTMA run by Kevin Murphy. Though he never spoke during an actual episode during the official seasons, it can be presumed that it is Cambot's voice heard during the "Robot Roll Call" portion of the opening theme from Season 1 through 5, and when shown during the opening of later seasons, though it is left ambiguous.

At the end of Season 7, Cambot was shown joining his fellow crew-members ascending into pure energy at the end of the universe. When the Satellite of Love crashed on Earth in the show's final episode, it is not specified whether Cambot survived the crash (although one could assume that he is the one filming the final scene). Cambot was not mentioned in *Mystery Science Theater 3000: The Movie*.

2.3 External links

- A page with instructions for building a Cambot (As seen in seasons 5-7)

- Forum for the discussion of building prop replicas of Cambot

Chapter 3

Cartoon Dump

Warning: Page using Template:Infobox television with unknown parameter "tv_com_id" (this message is shown only in preview).
Warning: Page using Template:Infobox television with unknown parameter "imdb_id" (this message is shown only in preview).
Warning: Page using Template:Infobox television with unknown parameter "asst_producer" (this message is shown only in preview).
Warning: Page using Template:Infobox television with unknown parameter "co-producer" (this message is shown only in preview).
Warning: Page using Template:Infobox television with unknown parameter "supervising_producer" (this message is shown only in preview).
Warning: Page using Template:Infobox television with unknown parameter "story_editor" (this message is shown only in preview).

Cartoon Dump is an online comedy series/video podcast created by Frank Conniff (formerly of *Mystery Science Theater 3000*) and animation historian Jerry Beck. A live version is currently making monthly performances at the Steve Allen Theater in Los Angeles, California and premiered in New York City in January 2008. The show is currently running on the first Mondays of each month at QED: A Place to Show and Tell in Astoria, New York, and hosted by Conniff.

3.1 Story and format

Set in a garbage dump, the show is a parody of stereotypical children's TV programming. The host of the show, "Compost Brite" (Erica Doering), despite being constantly cheery, obviously suffers from depression as well as anorexia nervosa. Compost Brite starts each episode with the show's theme song, a jolly tune played on a theatre organ (however, Compost Brite is seen playing a guitar). Each episode features slightly different lyrics. After this,

Compost Brite usually has a brief satirical discussion with the viewers on topics such as nutrition, dry heaves or how mediocrity pays off in the end. Later one of Compost Brite's friends, usually "Moodsy the Clinically Depressed Owl" (Conniff) joins in to sing a song or plug some fictional product which is sponsoring the show. This leads into the bulk of each episode, an obscure cartoon from Jerry Beck's archives.

3.2 Episodes

As of November 2007, there have been six episodes of Cartoon Dump. As the episodes have no official titles, they are shown here with the titles of their respective cartoons.

1. "Mighty Mister Titan"

2. "The Big World of Little Adam"

3. Bucky and Pepito: "The Vexin' Texan"

4. Captain Fathom: "Rustlers of the Sea Range"

5. "The Adventures of Spunky and Tadpole"

6. "The Adventures of Sir Gee Whiz on the Other Side of the Moon"

Chapter 4

Cinematic Titanic

Warning: Page using Template:Infobox film with unknown parameter "format" (this message is shown only in preview).

Cinematic Titanic is a project by *Mystery Science Theater 3000* (*MST3K*) creator and original host, Joel Hodgson.[1] The project involves "riffing" B-movies, in a manner similar to that of *MST3K*.[2] Joining Hodgson are many of the original *MST3K* cast, as well as some cast members who joined later in the show's run.[3] These include Trace Beaulieu, J. Elvis Weinstein, Frank Conniff and Mary Jo Pehl.[4] It was first performed live on December 7, 2007 and first aired on December 22, 2007.

On February 16, 2013, it was announced that the touring portion of Cinematic Titanic was going on an indefinite hiatus. According to an email sent out to members of the site, due to "5 people living in 5 different cities with different lives and projects, it has become increasingly difficult to coordinate our schedules and give Cinematic Titanic the attention it requires to keep growing as a creative enterprise and a business."[5] The final tour began on September 23, 2013.[6]

The cast of Cinematic Titanic at a live show

4.1 Description

Like *Mystery Science Theater 3000*, the series uses black silhouettes of the riffers placed over the films, but in the case of *Cinematic Titanic* they sit on both sides of the screen rather than just on the lower right.[7] Visual gags are frequent (such as Beaulieu's use of a cherry picker in *The Oozing Skull*), and there are two or three host segments per episode, all performed in silhouette.

4.1.1 Plot

The actors essentially play themselves as they participate in an experiment for some unknown (possibly shadowy) corporation or military force. The story currently provided to the cast is that there is a tear in the "electron scaffolding" that threatens all digital media in the world.[8] Their experience doing *MST3K* is key to the organization's plans. The riffing for each film is recorded to a "nanotated disc" and inserted into a "Time Tube" by Hodgson that descends into the frame at the end of every episode. The unknown organization is very firm on keeping the cast focused on their duties, providing no time frame for completion and requiring them to stay within the facilities at all times. They apparently have massive resources and an autonomous military force, which they use to keep the cast in line. As of now, the cast is inquisitive of the true purpose of the experiments but have no major problems as, aside from having to watch bad movies, they are well-treated.

4.1.2 Relation to *RiffTrax*

Michael J. Nelson, who took over as the star of *MST3K* upon Hodgson's departure from the show, has his own movie-riff series, titled *RiffTrax*. Nelson's project produces riffs for a wide variety of films, including many current and well-known movies, such as *Twilight*, *The Matrix* and *Lord of the Rings*.[3] He then posts the audio for sale on the *Riff-Trax* website.

The fact that *Cinematic Titanic* involves almost every *MST3K* writer and performer aside from Nelson, Kevin Murphy, and Bill Corbett, who happen to be the regular cast of *RiffTrax*, has prompted fan speculation about a rivalry between Hodgson and Nelson surrounding the two projects, but the pair have consistently denied that such a rift exists and expressed praise for each other's projects, pointing out that they fill different niches and there is more than enough room for both of them. Speaking with *Associated Content* in 2009, Mary Jo Pehl, who has worked in both projects, denied knowledge of any animosity between the *Cinematic Titanic* and *RiffTrax* groups. "I have no idea if there's some sort of family feud between the Joel vs. Mike factions. If there is, people ought to find better things to do with their time, like debating which way the toilet paper should hang on the spindle, or if the opening of pillow cases should face the outside of the bed or the inside when placed. *RT* does their thing, and *Cinematic Titanic* does theirs."

When asked about potential collaborations with those involved in *RiffTrax*, Hodgson told *New York* magazine, "I don't know. I think those guys — Bill, Kevin and Mike — are really talented, obviously. I think anything's possible, but I thought it might get confusing to try to merge them together or do crossover projects. I would never rule it out because it's all kind of the same universe. But *RiffTrax*, the idea of riffing on topical movies, is a different thing. And I like that the movie-riffing universe got bigger when they decided to do that, but we just do weird movies you've never seen before."[9]

4.1.3   Studio / "Live" DVDs

The cast riff on the movie East Meets Watts *at a live show in NYC*

In a question and answer session at the Tivoli Theatre in St. Louis, Missouri, it was announced that *Cinematic Titanic* would begin to release recordings from their live shows as

"Live" DVDs in an effort to bring the energy of their on stage antics into people's living rooms and further promote the stage show. When asked if this meant the demise of the studio produced DVDs, J. Elvis Weinstein said, "No, studio releases will return at some point in the future." The first of these "Live" DVDs to be released was *East Meets Watts*, which was recorded in front of a live audience during one of the group's performances in Los Angeles.

4.2   Releases

4.2.1   Release history

The first live performance was a private show for employees of Industrial Light & Magic on December 7, 2007.[4] After the live show, the cast reworked some jokes, delaying the original December 10 release date. The first episode of *Cinematic Titanic* was released on DVD to the public at midnight on December 21, via the download-to-burn company EZTakes. According to the *Cinematic Titanic* website, due to rights issues, the episode was not available for download until April 2, 2008.

Both the private show and the first release feature the B-movie *Brain of Blood*.[7] One of the original film's producers, concerned that creating multiple versions of the film could create marketplace confusion, requested that *Cinematic Titanic*'s version have a different name. To alleviate his concerns, *Cinematic Titanic* retitled their release *The Oozing Skull*.[10]

In June 2012, the first ten DVDs were made available for free viewing on Hulu.

In March 2013, Cinematic Titanic sold the last of their on-hand DVD stock and ceased pressing their own discs. From then on all releases were either in digital format, or through Amazon.com's print-on-demand service.[11]

4.2.2   Release list

Releases have been available to purchase as a physical DVD, and also as a download and burned DVD version.

4.3   List of live shows

The following is an incomplete list of live performances by Cinematic Titanic.

4.4 See also

- RiffTrax
- The Film Crew
- *Mystery Science Theater 3000*

4.5 References

[1] Meyer, John P. (January 12, 2008). "Good news for bad movie (and good humor) lovers: The MST3K crew are at it again". *Pegasus News*. Retrieved 2008-06-15.

[2] Ellis, Mary Beth (November 18, 2007). "Ex 'MST3K' stars, writers fill hole left by show". MSNBC. Retrieved 2008-06-15.

[3] Hoevel, Ann (September 6, 2010). "Cinematic Titanic's struggle with 'Weisenheimer's'". CNN. Retrieved September 6, 2010.

[4] Hodgson, Joel. "Greetings Friends". Cinematic Titanic. Archived from the original on 2007-11-19. Retrieved 2007-12-02.

[5] Beaulieu, Trace (February 16, 2013). "Cinematc Titanic will soon be riffing off". Twitter. Retrieved 2013-02-16.

[6] "Cinematic Titanic - The Masters of Movie Riffing - Live Tour". Cinematic Titanic. September 20, 2013. Retrieved 2013-09-20.

[7] Lewinski, John Scott (December 7, 2007). "Cinematic Titanic Steams Into Mystery Science Theater Waters". *Wired*. Retrieved 2007-12-15.

[8] Ryan, Maureene (December 11, 2008). "Joel from 'Mystery Science Theater 3000' is back with a new cinematic experience". *Chicago Tribune*. Retrieved 2008-12-11.

[9] Joseph Brannigan Lynch (April 16, 2010). "Mystery Science Theater 3000's Joel Hodgson on His Second Go-Round As a Movie-Mocker". New York Magazine. Retrieved April 29, 2010.

[10] Hodgson, Joel. "It's on!". Cinematic Titanic. Retrieved 2007-12-21.

[11]

[12] "Cinematic Titanic Episode Two Promo - Doomsday Machine" (video). *YouTube*. Cinematic Titanic. 22 May 2008. Retrieved 2008-06-15.

[13] Weinstein, J. Elvis. "There's A Lot of Buzz Around Here". Cinematic Titanic. Retrieved 2008-08-07.

[14] "Live Cinematic Titanic DVD Coming; New Tour Dates Announced". *Satellite News*. 2009-12-09. External link in |work= (help)

[15] "Next from Cinematic Titanic…". *Satellite News*. 2010-02-16. External link in |work= (help)

[16] "Show Guide - Danger on Tiki Island". *Cinematic Titanic*. Retrieved 2010-05-19.

[17] "Live Tour". *Cinematic Titanic*. Retrieved 2011-08-16.

4.6 External links

- Cinematic Titanic offline
- Official Youtube
- MST3K.info for Cinematic Titanic
- Interview with Joel Hodgson at IFC.com
- Interview at StarWars.com
- Review of Cinematic Titanic and its first episode
- DVD Talk discusses the first *Cinematic Titanic* episode

Chapter 5

Clowns in the Sky

Clowns in the Sky is the title of a CD featuring music from the first seven seasons of Mystery Science Theater 3000. Released in 1996 exclusively through the MST3K Info Club, the songs featured include the many versions of the opening theme, the closing theme, and various tunes that were featured in host segments throughout the show. The CD takes its name from the song featured in the last host segment of the episode Pod People, A Clown in the Sky.[1]

5.1 Tracks Listing[2][3]

5.2 References

[1] http://www.therobotsvoice.com/2010/12/the_13_best_mystery_science_theater_3000_songs.php

[2] https://itunes.apple.com/us/album/clowns-in-the-sky-i-ii/id337424254

[3] http://www.mst3kinfo.com/ward_e/listclown.html

5.3 External links

- http://www.therobotsvoice.com/2010/12/the_13_best_mystery_science_theater_3000_songs.ph

- hCttps://itunes.apple.com/us/album/clowns-in-the-sky-i-ii/id337424254

- http://www.mst3kinfo.com/ward_e/listclown.html

Chapter 6

Crow T. Robot

Crow T. Robot is a fictional character from the American science fiction comedy television series *Mystery Science Theater 3000* (*MST3K*). Crow is a robot, who, along with others, ridicules poor-quality B movies.

6.1 Overview

According to the *MST3K* storyline, Crow — like his fellow robots Tom Servo, Gypsy, and Cambot — was built by Joel Robinson, who created them to help him withstand the torment of watching bad movies on the Satellite of Love. On the Satellite, Crow was forced with the rest of the crew to watch horrible B-movies sent by mad scientist Dr. Clayton Forrester and his assistants. In episode 814, "Riding with Death", Crow describes himself as being made from molybdenum.

Crow's middle initial stands for "The". In episode #K19: *Hangar 18*, Joel stated that "Crow" was an acronym for "**Cybernetic Remotely Operated Woman**", giving Crow a brief identity crisis until Joel revealed he built Crow specifically to play this joke on him. Crow is also sometimes called "Art", primarily by late-series antagonist Pearl Forrester. This arose from a gag in episode 203 *Jungle Goddess* following a skit centered on the sitcom *The Honeymooners*, where Joel referred to Crow as "Art Crow" (in reference to "Honeymooners" co-star Art Carney). After Best Brains received a letter from a child who had evidently missed the cultural reference and labeled a drawing of Crow as "Art", the show's writers turned the name into a recurring joke.[1]

Crow was voiced by Trace Beaulieu from the beginning of the series through the end of season seven, and Bill Corbett from the eighth season until the end of the show. Corbett's Crow was noticeably more irritable, bitter, and impatient with the movies than Beaulieu's Crow had been. He will be voiced by Hampton Yount in the 2016 series revival.[2]

Crow's accomplishments during the show's run include:

- Writing several screenplays, including *Earth vs. Soup*

(his seminal work) (seen in *Earth vs the Spider*), *Peter Graves at the University of Minnesota* (*Beginning of the End*), *The Spy Who Hugged Me* (*Secret Agent Super Dragon*), and *Chocolate Jones and the Temple of Funk* (*Angels Revenge*). He also wrote a rather poorly researched documentary titled *Crow T. Robot's Bram Stoker's The Civil War*, and created another called *Let's Talk Women!*, in which he denies the existence of women. He also wrote a one man show titled "Robot on the Run".

- Being an avid member of the Kim Cattrall and Estelle Winwood fan clubs.

- Co-writing a satirical musical called *Supercalifragilisticexpiali-wacky!*

- During the Christmas episode #321, *Santa Claus Conquers the Martians*, he wrote a Christmas carol titled "Let's Have a Patrick Swayze Christmas", inspired by his favorite movie, *Road House*.

- Though all the SOL prisoners make it their goal to escape, the cause is almost always desire for freedom, not a result of the slew of B-movies. But, in episode #903, *The Pumaman*, Crow actually succumbed to the Mads' experiment. He decided this film had finally pushed him over the edge and that he could no longer take all the movies. He attempted to run away, hoping to find a satellite where he would be forced to watch *good* movies, but gave up on his escape attempt after mere seconds.

- In between segments of the movie Werewolf he successfully turns Mike into a Werecrow.

In the earlier seasons of the show, he usually announcing "Ladies and gentlemen, *Topo Gigio!*" in the voice of Ed Sullivan.

During episode 416, *Fire Maidens of Outer Space*, Crow acquired a double named "Timmy", to whom the trio quickly took a liking. (Timmy was actually the black-painted Crow used for the Shadowrama in the theater.) However, the double began playing tricks on Tom Servo and Joel, who blamed Crow for the actions. He eventually joined them in the theater during the movie and attacked Tom, cocooning him in a material identical to the xenomorph in *Alien*, forcing Joel (who referenced *Aliens* by growling, "Let go of him, you **bitch**!") and Crow to stuff him into an airlock and blast him out of the ship and into space. Timmy's appearance was a parody of the evil twin concept.

Crow holds the distinction of being the only SOL robot who ever visited Deep 13 (although Tom Servo and Gypsy once visited the alternate-earth version and Cambot is connected to the Mads' camera): In episode 615, *Kitten with a Whip*, he slid down the Umbilicus in an attempt to bring the SOL back to Earth, only to be frightened back up by Dr. Forrester (who then had Frank put a giant mousetrap beneath the Umbilicus). In the "Turkey Day" version of episode 701, *Night of the Blood Beast*, Crow attended the Forrester Thanksgiving dinner alongside such guests as Mr. B Natural, Pitch the Devil, and others. During this episode, Crow seems to have a friendship with Pearl Forrester, who seems to enjoy his company and the fact he would listen to her complain about her inept son, Dr. Forrester. Due to a time travel paradox in episode 821, *Time Chasers*, a second Crow lives in Minnesota, working at a cheese factory.

In the theatrical film based on the series, Crow distinguishes himself early on by attempting to tunnel back to Earth using a pickaxe. Even he admits the faulty logic employed in this scheme when he examines his calculations: "Well, look at that! 'Breach hull, all die' — I even had it underlined!" / "Well, believe me, Mike, I calculated the odds of this succeeding versus the odds I was doing something incredibly stupid, and...I went ahead anyway."

A running joke of Crow's character throughout the series, particularly in the last few seasons of the Sci-Fi channel's running of the show, is his frequent costume changes into film characters during the host segments. Often Crow has been known to take on the appearance and sometimes the personality of one of the characters in the movie that is currently being shown, usually ending with Mike, Tom Servo, or Joel using Crow's role-playing as a plot device or setup to a joke that mocks the film even more. This continuing effect finally was recognized by Crow himself during Episode 902, *The Phantom Planet*. At the conclusion of the film, Crow encounters Tom and Mike dressed as a 'Solarite' (one of the cheesy alien monsters from the film) and, having no memory of ever making or even deciding to make the costume, asks them "Have you guys ever noticed how I will see a movie, snap, then suddenly pretend I'm one of the characters in the movie and run about the ship?"

Another recurring joke that was more prominent during the earlier days of the show - dating, in fact, from the KTMA run - was Crow's apparent inability to distinguish between different types of animal; he would joyfully cry "Kitty!" when an animal appeared on screen, regardless of whether it was at all feline, and regardless of whether it was cute enough to elicit such a positive response.

6.2   Behind the scenes

Crow is a gold-colored puppet composed of, among other things, a soap dish eye cowl, ping pong ball eyes, a split plastic bowling pin mouth, a lacrosse face mask webbing, and Tupperware panels for the body. The original puppet for Crow was built by Joel Hodgson in a single night before filming the pilot episode. At the beginning of Season One, the puppet was redesigned and built by Trace Beaulieu, adding a second Tupperware tray to Crow's torso as well as movable eyes. Another version of Crow is used for the theater segments. This version is painted flat black. Of all the bots, Crow was the least changed from his KTMA incarnation to when the series became nationally broadcast. He's the only one who retained his general primary color.

Trace Beaulieu operated Crow in the initial KTMA season and throughout the Comedy Channel/Comedy Central years of seasons 1-7, as well as in *Mystery Science Theater 3000: The Movie*. During KTMA and season 1, Trace slowly refined Crow's voice. Originally, it was somewhat babyish, with a pronounced Minnesota accent. By season two, Crow's voice had become more sharp. Upon Trace's departure, Brooklyn-raised Bill Corbett took over Crow's operation. Corbett jokingly mentioned that during his time alone on the SOL Crow suffered a stroke, thus explaining the change of his voice and his less-than-graceful handling of the puppet.[3] Despite initial concerns from the fanbase regarding Beaulieu's departure, the reception for Corbett's performance was very positive and he quickly became a fan favorite. The show's writers later made a joke of the change in episode 904, *Werewolf*, by suggesting that Crow's inherent characteristics included a change of voice every seven years. When Joel Hodgson returned for a guest spot in episode 1001, *Soultaker*, he offhandedly suggests that Crow "changed his bowling pin" (that is, his mouth). From Season 8 onward, in the opening theme during the "Robot Roll Call", Crow can be heard exclaiming "I'm different!"

Many first-time viewers of the series are confused by Crow's appearance during the movie segments. Only the outline of his head can be seen, and (due to multistable perception) it can appear as if Crow is facing *toward* the viewer. This phenomenon was addressed in *The Mystery Science Theater 3000 Amazing Colossal Episode Guide* with illustrations comparing Crow to a Necker cube.[4]

At Dragon Con in 2009, Beaulieu and Corbett made a joint

appearance for a "Crow vs. Crow" panel discussion, in which they discussed their respective work with the character. The discussion is included as a bonus feature on the *Mystery Science Theater 3000: Vol. XX* DVD set from Shout! Factory.

6.3 Appearances in other media

- An episode of *Futurama* entitled *Raging Bender* has the gang visiting the theater, where Fry mockingly riffs on a newsreel intro before being shushed by the silhouette of a rather testy Crow-like robot ironically saying "Don't talk during the movie!"; beside him is a Tom Servo looking robot.[5]

- In the Archie Comics series *Sonic the Hedgehog*, issue #52, Sonic is sent into a 1920s variation of Mobius. In searching for the handheld computer Nicole, Sonic does battle with a number of robots, three of them resembling Crow T. Robot, Tom Servo and Cambot.

- In an issue of *Star Wars Tales*, Crow and Tom (with his cylinder head) are seen in the foreground of a comedic tale written by Peter David, starring a perky Force-using droid. He also appears in *Tag and Bink: Revenge of the Clone Menace*, along with Tom Servo and Gypsy.

- The June 8 2007 edition of the Cat And Girl comic features Crow as the President of South Vietnam.

- A Crow-like robot is seen among the dead on J'asik, the fourth moon of Kol, by the Autobot-Decepticon Alliance in *Transformers: Generation 2 #9*.

- Trace Beaulieu reprised his role as Crow for a brief cameo appearance in two episodes of the fourth season of *Arrested Development*, along with Joel.[6]

6.4 Crow Syndrome

The "Crow Syndrome" is a frequent joke on the show and MSTings, wherein Crow chimes in with an off-topic and/or excessively lewd comment and the other two reprimand him, often bemusedly and perturbedly shouting "Crow!" in response. "Crow Syndrome" is a general term, and is used in MSTings that do not feature Crow or other regular characters.[7]

6.5 References

- MST3K FAQ: What is this MST3K Thing, Anyway?

- MST3K FAQ: "This is my Bot! There Are Many Others Like It..."

[1] "Subtleties, Obscurities, Odds And Ends". *Mystery Science Theater 3000: Frequently Asked Questions*. Satellite News. Retrieved 2007-09-15.

[2] "Felicia Day, Baron Vaughn, and Hampton Yount Join the 'MST3K' Reboot". http://splitsider.com/. Retrieved 2015-03-12. External link in |publisher= (help)

[3] "Episodes: #801: The *[sic]* Revenge of the Creature". *CastleForrester.com*. Retrieved 2007-09-15.

[4] Trace Beaulieu ... (1996). "The Mystery Science Theater 3000 Amazing Colossal Episode Guide". Bantam: 159. ISBN 0-553-37783-3.

[5] *Futurama*, episode "Raging Bender" [2.21], 27 February 2000

[6] MST3K Pops Up in Arrested Development

[7] Web Site Number 9 MSTing FAQ, question 3.4

6.6 External links

- Crow T. Robot at the Internet Movie Database

- A page with instructions for building Crow

- Parts list for the above link

- Details of Crow T. Robot's construction through the entire run of the series.

- Forum for the discussion of building prop replicas of Crow T. Robot.

Chapter 7

Darkstar: The Interactive Movie

Darkstar: The Interactive Movie is an interactive movie video game written, produced, edited, animated, and directed by J. Allen Williams, owner of the American animation studio Parallax Studio. It starred the actor Clive Robertson[1] the and the original cast of the comedy series *Mystery Science Theater 3000* (Trace Beaulieu, Frank Conniff, Joel Hodgson, Mary Jo Pehl, and J. Elvis Weinstein).[2] The game also featured animations by the comic book artist Richard Corben[3] and was the final work of actor Peter Graves, who narrated the game. *Darkstar* was released online on November 5, 2010 through the company website and as a downloadable through Strategy First. It was re-released in stores December 9, 2011 in the United Kingdom and Ireland through Lace Mamba Global.[4]

7.1 Interactive movie

Darkstar differs from the standard game format in that it contains over thirteen hours of live action cinema–far more than any previous full motion video game. Creator and Parallax Studio CEO J. Allen Williams gave a brief description of the project to the website Slightly Deranged saying:

7.2 Plot

The player is Captain John O'Neil of the *Westwick*. You awaken from a cryogenic sleep that has spanned a period of over 300 years. As a result of the abnormally long hibernation, you have no memory as to who you are, where you are, or why you are there. Beside you are three other sleep chambers. One is empty, the other is occupied by a beautiful woman, and in the final chamber lies the body of a man–300 years dead and missing his left hand.

Your ship is damaged and helplessly adrift in the orbit of the ominous Theta Alpha III. An unknown crew member has deleted any data that provides an explanation as to why this is.

You know, despite the emptiness and desolation, that someone is in the cold silence waiting for you.

And, as if the day isn't bad enough, the Earth has been destroyed for three centuries.[6]

7.3 Cast

Darkstar has roughly thirteen hours of live-action cinema including a cast of all real actors. It stars Clive Robertson as Westwick Captain John O'Neil.[7] It also features the entire original cast of *Mystery Science Theater 3000* including its creator Joel Hodgson as Scythe Commander Kane Cooper, Trace Beaulieu as Westwick First Officer Ross Perryman, Frank Conniff as both Westwick Navigator Alan Burk and the voice of the quirky robot SIMON (Semi Intelligent Motorized Observation Network), Mary Jo Pehl as both Bridgebuilder Captain Beth Ingram and the voice of the computer Westwick Main, and J. Elvis Weinstein as Galactic Discovery II Captain Cedrick Stone. Also from *MST3K* is Beth "Beez" McKeever as the Westwick Pilot Paige Palmer who stars across from Clive Robertson. *Darkstar* was also the final work of the actor Peter Graves.[8]

In addition to Clive Robertson and the *Mystery Science Theater 3000* players, the production has a cast of nearly fifty actors, almost all of whom are local to Springfield, Missouri and the surrounding area.[9]

- Trace Beaulieu as Westwick First Officer Ross Perryman
- Bill Brown as B170 Pilot Billy Bob Brown
- Alan Bryce as President Timothy Brisbane
- Libby Chappell as F88 Pilot Lisa Hicks
- Frank Conniff as Westwick Navigator Alan Burk and the voice of SIMON (Semi Intelligent Motorized Observation Network)
- Valli Flores as Northstar Pilot Zoe Palmer

- Peter Graves as Narrator

- Lisa Hamaker as Solar Patrol Ship Pilot Landry

- Obie Harrup III as Captain of the Solar Patrol Ship

- Brad Hedrick as Bridgebuilder First Officer McIntyre

- Joel Hodgson as Scythe Commander Kane Cooper

- Brian McElroy as F88 Wing Commander B. Mitchell

- Beez McKeever as Westwick Pilot Paige Palmer

- Paul Oakley as Purgatory II Warden Tom Carlson

- Mary Jo Pehl as Bridgebuilder Captain Beth Ingram and the voice of Westwick Main

- Clive Robertson as Westwick Captain John O'Neil

- Todd Smith as B170 Pilot Rock Carnage (Francis Lieberwitz)

- Woody P. Snow as TurboTwin Pilot Chuck Nordstrum

- Aaron Wahlquist as F88 Pilot Stewart

- J. Elvis Weinstein as Galactic Discover II Captain Cedrick Stone

- J. Allen Williams as B170 Pilot Dean Wilder

- Margaret Noel Williams as the voice of MAGS (Motorized Automated Girl for SIMON)

7.4 Production

Darkstar was written, produced, animated, edited, and directed by J. Allen Williams over the course of nearly a decade. Though most of the credit goes to Williams alone, a number of others contributed to the enormous production. Additional animations for *Darkstar* were done by the prolific American illustrator and comic book artist Richard Corben who is best known for his "Den" character featured in the 1981 film *Heavy Metal* and for his comics featured in the magazine of the same name. Other noteworthy crew members include cinematographer Roger Jared, co-producer Mark L. Walters, electronic media producer Dahlia Clark, and composers Jimmy Pitts, Bill Bruce, and J. Allen Williams.[10]

7.5 Soundtrack

The soundtrack to *Darkstar* was composed and performed by Jimmy Pitts (keyboards and pianos), Bill Bruce (guitars and percussion), and J. Allen Williams (bass guitar) under the moniker "Progressive Sound And MetalWorx". Two other performers include Brent Frazier (guitars) and James Lee Dillard (percussion).[11] Additional music for was composed by Ruell Chappell, an original member of the Ozark Mountain Daredevils. Though the soundtrack was originally intended to feature over an hour of music by the rock band Rush, negotiations with Universal Music eventually dissolved and Williams was forced to replace much of the footage with an entirely original score.[12]

The soundtrack was released in tandem with the game and features 38 tracks of music on a 2-disk set.[13]

7.5.1 Disk One

1. "Surfing the Velvet Abyss" (3:17) - by Bill Bruce

2. "Mindscape" (3:15) - by Jimmy Pitts, Ruell Chappell

3. "Mirrors Tell a Different Tale" (3:05) - by Jimmy Pitts

4. "Systematic Overload" (3:50) - by Bill Bruce

5. "Solar Wind Chimes" (1:58) - by Jimmy Pitts, Ruell Chappell

6. "Lost in Eternity (Paige's Theme)" (4:05) - by Jimmy Pitts

7. "An Insignificant, Distant Nova" (2:17) - by Jimmy Pitts

8. "FDS2" (1:49) - by Bill Bruce

9. "Dementia in Absentia" (3:52) - by Jimmy Pitts

10. "Deceptive & Dangerous" (4:15)- by Bill Bruce

11. "Noir in A Minor" (3:11) - by Jimmy Pitts

12. "Considered Gone" (1:34) - by Bill Bruce

13. "Light Years of Darkness" (2:11) - by Jimmy Pitts

14. "Requiem for a Blue Dot" (3:44) - by Jimmy Pitts

15. "147" (4:10) - by Bill Bruce

16. "Winds of Fate" (1:42) - by Jimmy Pitts, Ruell Chappell

17. "Theoretical Paranoia" (2:18) - by Jimmy Pitts

18. "Falling Down Stairs" (4:22) - by Bill Bruce

19. "A Little Dream All My Own" (3:32) - by Jimmy Pitts

20. "Winds of Fate (Reprise)" (4:41) - by Jimmy Pitts, Ruell Chappell

21. "Underneath" (4:06) - by Bill Bruce

22. "Light Years of Darkness (Reprise)" (2:14) - by Jimmy Pitts

23. "Another Time and Place" (3:06) - by Bill Bruce

7.5.2 Disk Two

1. "Bonechipper" (5:21) - by Bill Bruce

2. "Eternal Twilight" (6:23) - by J. Allen Williams

3. "Day of Darkness" (3:46) - by J. Allen Williams

4. "Tomorrows Children" (6:00) - by J. Allen Williams

5. "Corridors of Time" (3:16) - by J. Allen Williams

6. "The Edge of Insanity" (4:36) - by J. Allen Williams

7. "Knights of Vengeance" (5:14) - by J. Allen Williams

8. "One Two One" (4:11) - by J. Allen Williams

9. "Guardian" (7:35) - by J. Allen Williams

10. "A Trinity of Sons" (4:36) - by J. Allen Williams

11. "The Edge of Nowhere" (2:52) - by Jimmy Pitts

12. "Holy War" (5:04) - by J. Allen Williams

13. "The Secret Sign" (4:53) - by J. Allen Williams

14. "Psychic Pilgrims" (4:42) - by J. Allen Williams

15. "Lasting Memory" (4:12) - by Bill Bruce

Reception for the soundtrack has been positive. In a review of *Darkstar* at diehardgamefan.com, Alex Lucard rated the soundtrack as "Unparalleled" and wrote, "This is without a doubt one of the the [*sic*] best scores I have heard all year."[14]

7.6 Reception

Darkstar: The Interactive Movie received a mixed to negative response from game critics. On the review aggregator Metacritic, the game has a weighted average of 36% indicating "generally unfavorable reviews."[15]

Adventure game reviewers generally praised the game. Drummond Doroski of Adventure Gamers gave it 3 out of 5 stars, writing, "It's not a game for everyone, as some are sure to be turned off by the rarity and simplicity of its puzzles, while others may not relish a return to the infamous days of live actors as their game characters, particularly when some of the acting reminds us why this isn't always a good idea." He concluded, "*Darkstar* may be light on actual gameplay, but it's rich in cinematic storytelling, and for many science fiction and FMV fans, that's sure to be more than enough."[16] J. Robinson Wheeler of the adventure game website *Brass Lantern* similarly described it as "challenging, fun, lovingly and painstakingly rendered and crafted, and worth playing."[17] Alex Lucard of *Die Hard Game Fan* highly praised the game, giving it a final score of "Very Good Game!" and writing, "When all is said and done, *Darkstar* is one of the best indie games I have ever played in my thirty years of gaming. It's one of the ten best games I have played in 2010. Most of all, it was well worth the wait and then some. Don't let this thing pass you by simply because it doesn't have a multi-million dollar ad budget. You've been warned."[14]

Conversely, Games TM gave the game 1/10, saying simply "It's not a game."[18] PC PowerPlay also reviewed it negatively, with their 2/10 review calling *Darkstar* "A crime. Send this one to the colonies."[19]

7.7 References

[1] "CliveRobertson.110.com".   Cliverobertson.110mb.com. Retrieved 2010-09-07.

[2] "MST3Kinfo.com". MST3Kinfo.com. Retrieved September 7, 2010.

[3] "CorbenStudios.com". CorbenStudios.com. 2010-05-21. Retrieved September 7, 2010.

[4] Lace Mamba will release Dark Star in December 2011 in UK. *GameBoomers*. Retrieved 30 January 2012.

[5] "Darkstar   Interview   www.Slightly-Deranged.com". Slightly-deranged.com. Retrieved 2010-09-07.

[6] "Summary - Darkstar Prospectus" (PDF). Retrieved 2010-09-07.

[7] "Clive Robertson Bio". cliverobertson.110.mb.com. Retrieved 2010-09-07.

[8] "Peter Grave on". Themoneytimes.com. 2010-03-15. Retrieved 2010-09-07.

[9] "" (2010-04-01). "Darkstar KY3 Interview on YouTube". Youtube.com. Retrieved 2010-09-07.

[10] "Official Darkstar". Darkstar.gs. Retrieved 2010-09-07.

[11] "Darkstar Soundtrack Sample". Youtube.com. Retrieved 2010-09-07.

[12] "DARKSTAR IS FINALLY HERE!". 2112.net. 22 November 2010. Retrieved 10 December 2012.

[13] "Soundtrack details at". Darkstarstore.com. Retrieved 2012-03-14.

[14] Lucard, Alex (November 18, 2010). "Review: Darkstar – The Interactive Movie: Captain's Box (PC)". *Die Hard Game Fan*. Retrieved June 12, 2016.

[15] "Darkstar: The Interactive Movie". Metacritic. Retrieved June 12, 2016.

[16] Doroski, Drummond (January 21, 2011). "DARKSTAR: The Interactive Movie REVIEW". *Adventure Gamers*. Retrieved June 12, 2016.

[17] Wheeler, J. Robinson (2010). "Darkstar Review". *Brass Lantern*. Retrieved June 12, 2016.

[18] "Review summary at Metacritic". metacritic.com. 2012-01-29. Retrieved 2014-02-24.

[19] "Review summary at Metacritic". metacritic.com. 2012-01-29. Retrieved 2014-02-24.

7.8 External links

- Official website

- *Darkstar: The Interactive Movie* at the Internet Movie Database

Chapter 8

Dr. Clayton Forrester (Mystery Science Theater 3000)

Dr. Clayton Deborah Susan Forrester is a fictional character on the television series *Mystery Science Theater 3000* (*MST3K*).

Named for the hero of the 1953 film *The War of the Worlds*, Dr. Forrester was the chief mad scientist on the show from its inception in 1988 through the seventh season in 1996, and also appeared in *Mystery Science Theater 3000: The Movie* in 1996. He was played by Trace Beaulieu.

8.1 Character background

Forrester originally worked with fellow "mad" Dr. Laurence Erhardt at the Gizmonic Institute, until they moved their operations to the bowels of Deep 13. He engineered the kidnapping of janitor Joel Robinson, shooting him into space on board the Satellite of Love.

According to the show's mythos, Dr. Forrester (full name: Clayton Deborah Susan Forrester, perhaps because his mother wanted a daughter; he has referred to himself as Clayton "Stonewall" Forrester and Clayton "Firebrand" Forrester) had been a mad scientist ever since his youth, when he was a member of Evilos (a mad scientist version of Webelos), where he grafted the rear end of a dog onto the rear end of a cat; he has traced his scientific ambitions back to "Oslo...I was found drunk and woozy...scratching the name "Paula Cranston" into my thigh with a nail". Other pivotal moments in his early life include a 1956 visit to "Sun Valley...[where] I was found behind the soft-serve machine, drooling over a picture of Dick Button" (Since Trace Beaulieu was born in 1958, Forrester must thus be at least a decade or so older than the actor who portrayed him.) and a visit to "the Ice Capades, [where] I was hot-riveting my kneecaps to Peggy Fleming's zamboni."

Forrester's high school career was typified by a series of humiliations, presumably contributing to his rather deranged personality. Frequently teased by classmates, he received a "shameful expulsion" from the Chess Club, suffered a "shameful shower incident" during his sophomore year, got rejected by the Swing Choir, was frequently victimized by book-dumpings after typing class, was forced to do power sit-ups in gym, and received "the revulsion, scorn, and rejection of all the pretty girls." At some point, he was struck by lightning, resulting in the white streak in his hair and mustache.

While earning his doctorate, Forrester took some undergraduate courses in Super-Villainry, and at some point he joined the Fraternal Order of Mad Science. He was a frequent attendee of the Mad Scientist Convention, although he lost the convention's invention contest each year (on one occasion his entry, "the More Painful Mouse Trap", was met only with laughter). In response to his rejections, he has blown up the convention center twice and once used incendiaries to not "actually make the building blow up, it just made it burn...really quickly".

While working at Gizmonic Institute, Forrester and his assistant, (Dr. Laurence Earhardt from K-01 to the end of Season One, at which point he was replaced by TV's Frank without explanation) sent Joel cheesy movies which he was forced to watch, in order to find a movie that would drive people mad and allow him to take over the world. In response, Joel built several robot friends to keep him company, and keep himself from being driven mad. Joel, Crow T. Robot, and Tom Servo mocked each of the movies they were forced to watch. During Joel's time on *MST3K*, Forrester participated in Invention Exchanges with Joel and the 'bots. He would show his invention, then Joel would show his invention.

At the end of the sixth season, Frank was assumed into Second Banana Heaven by the angel Torgo the White, an event that, surprisingly, deeply saddened Forrester, reacting as though he had lost his best friend, even lamenting Frank's loss with the song *"Who Will I Kill?"*. In the seventh season,

Pearl Forrester (Mary Jo Pehl) joined her son Clayton to help him out. When Trace Beaulieu left the series, she took over as the head mad, and continued sending bad movies to hapless temp Mike Nelson (played by Michael J. Nelson, who had replaced Joel halfway through the 5th season) and the 'bots.

8.2 Demise

Dr. Forrester detaches the Satellite of Love from Deep 13.

Clayton's last appearance was *Laserblast* (Episode #706), where he announces that his funding has been cut, causing him to pack up Deep 13 and cut loose the Satellite of Love. The end of the episode is a parody of *2001: A Space Odyssey*, in which an old Clayton tries to reach a Monolith-like giant videocassette labeled "The Worst Film Ever Made". In the final scene, he is reborn as a star child. When Pearl muses about another chance to raise her son, he utters his final words of the Comedy Central series: "Oh, poopie." When he was leaving the show Trace Beaulieu said "It's kind of bittersweet, but it was time for me to go,".[1]

When the show moved to the Sci Fi Channel for its eighth season and the setting changed to the future year 2525, it was revealed that although Pearl had intended to do a better job of raising Clayton the second time around, she had somehow never actually gotten around to doing so and he had grown into much the same sort of adult he was before. Pearl therefore killed him by smothering him with a pillow. Oddly enough, although Pearl would have murdered Dr. Forrester some forty years to fifty years after the events of Episode #706, when the SOL crew returned to the 20th century, it was at the same time they had left it, i.e. at the end of Episode #706, which would mean that a second Pearl and a baby Clayton were still alive at that time. However, because MST3K is "just a show" which advises its viewers to "really just relax", this was never addressed.

In 2008, Trace Beaulieu, Joel Hodgson and Frank Conniff reprised their characters for a brief DVD skit explaining the rights issues for one of the films. The skit depicted Joel and the 'bots back on the ship and Forrester (as an adult) back at the lab with Frank. It is not known where (or if) the scene fits within the show's continuity.

8.3 References

[1] "Mad scientist Beaulieu retires". *The Vancouver Sun*. Jul 27, 1996. Retrieved 25 September 2013.

8.4 External links

- Dr. Clayton Forrester at the Internet Movie Database

Chapter 9

Dr. Laurence Erhardt

Dr. Laurence "Larry" Erhardt is a fictional character and one of the two original villains on the cult television show *Mystery Science Theater 3000*. He was played by Josh "J. Elvis" Weinstein.

9.1 Role

Dr. Erhardt is a mad scientist and was Dr. Clayton Forrester's first assistant. He is a very different character from Forrester's later assistant, TV's Frank. He is more high-spirited than Frank, speaks with a high, squeaky voice, has curly black hair, and thick glasses. Erhardt is himself a "mad scientist" like Forrester, whereas Frank is neither a scientist nor, for the most part, mad. Just like Frank, however, Erhardt goes through many tortures by Dr. Forrester. His catchphrase was to, after soberly describing a horrible movie's plot or some other dire scenario, cheerily proclaim "*Enjoy!*"

The Mads perform a commercial for their fictitious drive-in restaurant, "Clay and Lar's Flesh Barn"

Little is known of Erhardt's past save that he became a "mad scientist" while working at a zoo; he went mad when, in his own words: "...they promised me students, but all I got were monkeys! Monkeys! Monkeys! So I took off my wetsuit, dropped that hedge clipper, and walked out of that zoo forever!" Any additional details are left for the viewers to fill in themselves since MST3K was, after all, "just a show", and the viewers should "really just relax".

Eventually finding employment at Gizmonic Institute, he became Dr. Forrester's assistant and worked with him to launch Joel Robinson into space, where the pair conducted their movie-watching experiments on him.

Dr. Erhardt was only on the show for the KTMA season and the nationally telecast Season 1. When Frank was first introduced on the show and Joel asked what happened to Dr. Erhardt, Frank simply held up a milk carton with Erhardt's picture on it saying, "He's missing". Since Best Brains, the production company behind MST3K, discouraged the replaying of Season 1 episodes and had no rights to the KTMA episodes, Erhardt fell into obscurity, since few fans had actually had a chance to see him. After the milk carton reference, he was mentioned only once more, in episode #313 - "Earth vs. the Spider" featuring a scene where a policeman closely resembling Erhardt is eaten by a giant spider, and Joel and the bots speculated that this was the true fate of Erhardt.

9.2 Behind the scenes

Dr. Forrester and Dr. Erhardt were first shown in episode K06, during the show's original season on KTMA TV23. During this season, and season 1, J. Elvis Weinstein (who was credited under his real name Josh Weinstein) held a dual role as both Dr. Erhardt and the voice of Tom Servo, and he also provided the voice of Gypsy during the KTMA years (replaced in season one by Jim Mallon). Weinstein, the youngest of the MST3K creators, left the show after its first nationally telecast season due to creative differences once the show began being scripted instead of being ad-libbed. His youth and relative inexperience was also said to have rubbed his older co-workers the wrong way. In an interview, he stated the show simply quit being fun for him when they moved from KTMA to The Comedy Channel because it became "a business".

9.3 External links

- Dr. Laurence Erhardt at the Internet Movie Database

Chapter 10

Edward the Less

The Adventures of Edward the Less, better known simply as *Edward the Less*, is a 2001 animated miniseries fantasy comedy created by the former cast of the popular show *Mystery Science Theater 3000* for SciFi.com, the Sci Fi Channel website. It tells the story of Edward, a Pudge who reluctantly embarks on a quest to destroy a magic token before it falls into the hands of the evil Dark Person.[1]

10.1 Plot

Edward the Less is a Pudge, a race of short and merry people who spend their days in the Pudgelands dancing, eating horrible food, singing inane songs and whistling. But Edward is frustrated with the overly merry lifestyle and dreams of seeking out a magical city island he has read about, which includes tall people, skyscrapers, pizza and people who ride around in yellow cars and push each other around. Edward decides to leave his hamlet of Pushington Downs, along with his faithful sidekick Soapwort McFuggletoes. During their journey, the duo come into possession of a magic token, which they are forced to destroy so it will not fall into the hands of the evil Dark Person.[1]

10.2 Cast

- **Bill Corbett** as Edward[2]

- **Kevin Murphy** as Soapwort McFuggletoes[2]

- **Michael J. Nelson** as The Noble One[2]

- **Paul Chaplin** as Primatene[2]

- **Mary Jo Pehl** as Ariadrina and The Lorekeeper[2]

- **Patrick Brantseg** as Walt[2]

- **Mike Dodge** as The Narrator[2]

10.3 Production

Following the cancellation of the cult comedy *Mystery Science Theater 3000*, performers and writers from the show Patrick Brantseg, Bill Corbett, Kevin Murphy, Michael J. Nelson and Paul Chaplin conceived and created *Edward the Less*. The miniseries is a parody of the fantasy genre in general, but most especially of J. R. R. Tolkien's *The Lord of the Rings*; *Edward the Less* was inspired in large part by the release of *The Lord of the Rings* film trilogy, the first film of which was also released in 2001. The story structure, as well as specific characters and elements of *Edward the Less*, are directly inspired by *The Lord of the Rings*, including the Pudges, which are spoofs of Hobbits, and the Dark Person, which is inspired by the Dark Lord Sauron.[1]

Michael Nelson created the music, much of which is comical music Nelson considers "a parody of real music."[1] Some of the music was composed on ACID Pro, a digital audio workstation by Sony Creative Software.[1] *Mystery Science Theater 3000* alum Mary Jo Pehl joined the creators in voicing the characters, and Mike Dodge, who served as an *MST3K* writer for a few seasons, was cast as the voice-over narrator. Kevin Murphy said producing and recording *Edward the Less* went very smoothly because the whole crew had worked together before and quickly found their old sense of comedic timing again.[1]

Artists Rich Larson and Steve Fastner, who have worked together on several works of fantasy art including comics and book cover illustrations, created the animated drawings for *Edward the Less*. Larson worked with the show creators on *Mystery Science Theater 3000: The Movie* and was looking forward to illustrating *Edward the Less* because he was often critical of *Lord of the Rings* illustrations and looked forward to spoofing them.[1]

Edward the Less was written in three parts, in part because fantasy epic stories are usually structured as trilogies similar to *The Lord of the Rings*. Kevin Murphy said it was also structured this way so that the story could be easily continued if SciFi.com wanted more episodes beyond the original

13 they ordered. Murphy felt *Edward the Less* could work well in other media, including a book or a radio series in the style of *The Hitchhiker's Guide to the Galaxy.*[1]

10.4 Notes and references

[1] "SCIFI.COM | Edward The Less | Video Interview". *Sci-Fi Channel.* Internet Archive: Sci-Fi Channel. 2001. Archived from the original on 20 December 2002. Retrieved 3 June 2013.

[2] "SCIFI.COM | Edward The Less | Cast and Crew". *Sci-Fi Channel.* Internet Archive: Sci-Fi Channel. 2001. Archived from the original on 13 March 2002. Retrieved 3 June 2013.

10.5 External links

- Official site on SciFi.com Archived

- *The Adventures of Edward the Less* at the Internet Movie Database

Chapter 11

Gypsy (Mystery Science Theater 3000)

Gypsy is one of the robot characters on the television series *Mystery Science Theater 3000*. She is larger and less talkative than the other robots. Gypsy normally only appeared during the show's host segments and introduction, but briefly took a seat in the theater to watch the movie in episode #412 (*Hercules and the Captive Women*). She was only able to deliver a couple of "riffs", and left after realizing how bad the movie was. Along with the other robots, Gypsy was designed and built by series creator Joel Hodgson. He named Gypsy after a pet turtle his brother once owned, as the robot's size and ponderousness reminded him of the turtle.[1]

11.1 Role

According to the *MST3K* storyline, Gypsy takes care of the higher functions on board the Satellite of Love. She needs to use most of her computing power to handle this responsibility, which generally makes her appear slow-witted when dealing with others. The episode #207: Wild Rebels briefly demonstrated a much brighter Gypsy when the demand on her systems was temporarily reduced. (She also made a quick appearance in the theater during this experiment when someone mentioned *Voyage to the Bottom of the Sea*, on account of her oft-stated fascination for that show's star, Richard Basehart.)

As the show progressed, she became a more frequent participant in the host segments, and appeared more intelligent, even attempting to sit in on an experiment during one of the Hercules movies, *Hercules and the Captive Women*. But she begged out after a few minutes when she "realized there were these things" she had to go take care of. Gypsy is a big fan of Richard Basehart; her brain was once X-rayed, and found to contain RAM chips and a photo of Richard Basehart. She is also a notary.

Gypsy played a key role in the show's overall plot in episode #512, *Mitchell*, when she overheard the mad scientists discussing killing Mike Nelson, who was currently working at

Gizmonic Institute as temp, and mistakenly believed they were plotting to kill Joel. Terrified, she worked to come up with a plan to help Joel escape. Talking with Mike over the viewscreen, they were able to locate an escape pod. At the end of the episode, Gypsy launched Joel off of the Satellite of Love in the pod, marking Joel's departure as a regular character. The mads then plotted to send Mike as a replacement. Mike became the regular host for the remainder of the episodes.

In season six, the show introduced a device usually referred to as the Umbilicus, although in some episodes it was referred to as either the Umbilicon or the Umbiliport. The Umbilicus was a long tether that connected the Satellite of Love to the underground lair Deep 13, and allowed objects to be sent back and forth between Mike and the mad scientists who had stranded him in orbit. In its first appearance on the show, in episode #601: *Girls Town*, the Umbilicus was directly connected to Gypsy's snake-like body, with objects being sent or received through her mouth. The receiving station was later changed to an oven-like hatch on the bridge of the Satellite of Love.

Gypsy also developed an independent Bohemian feminist side to her personality, beginning with the *Creeping Terror* episode during the Mad Scientist's "coffeehouse poseur" experiment, when she sang, "You, the middle-class white male, are my personal oppressor!" Eventually, as the series went on, she penned her own cabaret-like one-woman show, "Gypsy Rose Me!" (a reference to burlesque performer Gypsy Rose Lee).

After the Satellite of Love crashed into the Earth in the final episode of MST3K (#1013: Diabolik), Gypsy is not present with Mike and the Bots in their shared apartment. Tom Servo is seen reviewing a "ConGypsCo Annual Report", and the guys reflect on their failure to have taken up new corporate mogul Gypsy on her public offering.

During KTMA episode #6, Gamera vs. Gaos, Joel claimed to have tried to program her voice to resemble Kim Carnes, but messed it up. In a sketch in episode #13, SST: Death Flight, Gypsy was uniquely voiced and operated by a woman

(Faye Burkholder). In episode #409, The Indestructible Man, Gypsy swapped voices with Magic Voice in the opening host segment.

In the Season 1 pilot, Gypsy managed to uncoil herself, revealing that she was at least 50 feet long. Joel explained that he just kept making her longer and longer compulsively when he built her (comparing it to, "When you start connecting paper clips, you get hooked on it...").

11.2 Behind the scenes

Gypsy's head was built out of a "Century Infant Love Seat". Parts of an Eveready flashlight were used for her eye, the white rubber "hood" portion generally absent after the first couple of seasons. Foam tubing was used on her lips (usually a dull light gray), and her neck was made of a long black PVC hose. The head portion was painted a metallic purple, with a small amount of metallic blue on the inside of her mouth. Occasionally the tubing of her lips is colored red, giving the effect that she's wearing lipstick.

During the initial KTMA season a different children's car seat was used for the head portion, the entire puppet painted a copper color. This Gypsy, with small alterations, would later be used for Cambot during the opening of season one episodes.

During the KTMA season, Gypsy's voice and operation were handled by Josh Weinstein. From the first to eighth nationally telecast seasons, Gypsy was voiced and handled by Jim Mallon, who was one of the producers and writers on the show. Starting midway through season 8 in episode #815, Agent for H.A.R.M., the job of operating and voicing Gypsy was handed over to Patrick Brantseg, who also performed most of the puppetry during filming of *Mystery Science Theater 3000: The Movie* so as to allow Jim Mallon to focus on directing rather than working the puppet. Patrick Brantseg would handle the role for the remainder of the series. For the new MST3K flash series on the show's official website, Jim Mallon once again provides the voice of Gypsy.

11.3 References

[1] "Satellite News - 20 Questions Only Joel Can Answer about MST3K". Mst3kinfo.com. Retrieved 2010-05-14.

11.4 External links

- A page with instructions for building a Gypsy

- Parts list for the above link

- Forum for the discussion of building prop replicas of Gypsy

Chapter 12

Joel Robinson

Joel Robinson is a fictional character featured in the American science fiction comedy television series *Mystery Science Theater 3000* (*MST3K*). He was portrayed by series creator Joel Hodgson.

12.1 Overview

The show's theme song explains Joel's backstory: Formerly a janitor and inventor for Gizmonic Institute, Joel was launched into space by his boss Dr. Clayton Forrester and co-worker Dr. Laurence Erhardt - later replaced by TV's Frank - as part of an experiment to see which bad movies were capable of destroying the human mind. Joel built the 'Bots Tom Servo, Crow, Gypsy, and Cambot to keep him company, but in doing so used parts that apparently caused him to lose the ability to control over when the films would stop and start. Though bombarded with many horrible films, he tends to take his captivity in benign stride, delivering most of his riffs in deadpan, holding no ill will against his captors and even affectionately calling them "the Mads" (among other amusing nicknames such as "the Overlords") while riffing on popular culture ("Auntie Em and Toto") or things found in Minnesota ("Milavetz and Associates", a prominent Twin Cities-area law firm).

As the opening theme song said, Joel generally wore a red jumpsuit during most of his time as host, but on occasion would wear other colors, such as tan (during the show's first improvisation season on KTMA) or green or bright aqua colored (often worn during Season 2). From season 2 episode 212 through his departure in episode 512, Joel wore a darker, maroon colored jumpsuit, though the original red jumpsuit (and second season green one) remained in the show's intro and opening theme.

Joel was the host from 1988-93. Episode #512, *Mitchell*, was his final episode as host; beginning with the following episode (#513 *The Brain That Wouldn't Die*) he was replaced by Mike Nelson, played by series head writer Michael J. Nelson.

During *Mitchell*, Gypsy overheard The Mads talking about eliminating Mike, the temp worker who was assisting them with an "evil-scientist audit". Hearing them refer to Mike as a "be-jumpsuited fool" and thinking they were actually talking about Joel, she struggled to think of some way to help Joel escape his alleged fate. Joel escaped the Satellite of Love (S.O.L.) in a previously undiscovered escape pod (named the *Deus Ex Machina*) mislabeled as a crate of "Hamdingers" that Mike told Gypsy about. Mike managed to override The Mads' control over the S.O.L. so that Gypsy could get Joel into the escape pod and launch it. After Joel's departure from the S.O.L., the Mads tried to rescue him, but gave up seconds later when they discovered that he had already safely landed in the Australian Outback. Instead of killing Mike, as they initially planned, they sent him up to take Joel's place. After Joel's departure he returned just once, in the show's final season (episode 1001: *Soultaker*), having turned the escape pod into a makeshift spaceship. He returned to the S.O.L. to make repairs to parts that were programmed to self-destruct a decade after the ship's launch into orbit and give Mike a pep talk, after which he left the satellite through the corridor to the theater, where he had entered it earlier in that episode. Joel reported that, since leaving the ship, he had traveled around the Australian outback, doing pyrotechnics for the band "Man or Astroman?", and that he was currently working as a manager at a Hot Fish Shop in Osseo, Minnesota. (In the real world, the most famous Hot Fish Shop in Minnesota had closed the weekend that episode 1001 aired, although the shop was located in Winona, Minnesota, rather than Osseo.)

After his departure, Joel was mostly forgotten until his appearance in *Soultaker*. However, he was indirectly mentioned in the episode *Santa Claus* when Gypsy gives Mike a sweater with the word "Joike" written on it, explaining that she had started it when "the other guy" was present, but finished it after Mike's arrival. He was also indirectly referenced in *Time Chasers* by Mike's brother Eddy (who, due to a time travel mistake caused by Crow, was now in Mike's place on the SOL), who called him the "sleepy-eyed guy". After his visit in *Soultaker*, he was mentioned again in episode 1004 (*Future War*) during the film's credits. Mike attempts to do something Joel-esque, worrying the Bots, and Crow eventually asks if this has anything to do with Joel stopping by recently. Joel was referenced again in episode 1008 (*Final Justice*), when Mike believed he could escape after "suffering though a horrible Joe Don Baker movie" like Joel had previously (the Bots revealed that Mike's escape pod was actually the ship's water heater).

Joel's tenure as host was marked by "invention exchanges", where Joel and his mad scientist tormentors would come up with wacky inventions in a contest with each other. These sketches were a good match for Hodgson, who began his career as a prop comic; indeed, many of the inventions were items originally found in his standup act. The gag remained during early episodes with the show's second host, but was quickly done away with (since the writers wanted to focus on Mike's strengths in portraying comic characters), as were any references to Gizmonic Institute, which Hodgson owned the rights to. The in-show reasoning behind the disposal of invention exchanges was that they were part of Gizmonic corporate culture, which Mike (having never worked at the Institute) knew nothing of. Another change was the relationship between host and bots; whereas Joel was more of a parental authority figure to Crow and Servo (in keeping with his status as their creator), the pair treated Mike more as a peer, occasionally subjecting him to pranks which they never would have considered playing on Joel.

Joel's "sleepy eyed" persona was reportedly the result of Hodgson staying up all night working non-stop on the pilot, which was kept after the pilot was shot because the crew and other performers thought it was funny. Off-camera, Hodgson wears eyeglasses, which can be occasionally seen when he turns his head in profile during scenes in the theater.

12.2   Name

In the pilot episode of *MST3K*, Hodgson simply used his real name. During its run on KTMA and the first season on the nationally-broadcast Comedy Channel, Joel's last name was never mentioned on air. During the second nationally-broadcast season, his character began using the surname Robinson, after the protagonist of *Lost in Space*.[1] (When Hodgson was later interviewed on *Space Ghost: Coast to Coast*, Space Ghost referred to him as "Mr. Lost in Space himself, Joel Robinson".)

12.3   References

[1] Jack Hagerty and Jon Rogers, *The Saucer Fleet* (Ontario: Apogee Books, 2008), 116, 278.

12.4   External links

- Joel Robinson at the Internet Movie Database

Chapter 13

List of Mystery Science Theater 3000 characters

Over its eleven-year run, *Mystery Science Theater 3000* saw the arrival and departure of various characters. In addition, it also featured many recurring guest characters. Below is a listing of both the main characters and the recurring characters from the series.

13.1 Main characters

Notes

1. ^ "Joel Hodgson" during season 0; Simply "Joel" (no last name) during Season 1.

2. ^ Simply "Frank" during seasons 2 and 3.

3. ^ Guest appearance only.

13.2 Recurring guest characters

- Jerry and Sylvia (various unpaid interns at Best Brains) - two "mole people" from the movie *The Mole People* (featured much later as a season 8 episode) who occasionally assisted the Mads and stopped by for social events. They also work Deep 13's camera in the first host segment of "Lost Continent". Presumably named after Sylvia and Gerry Anderson, the creative team behind *Space: 1999* and Supermarionation shows like *Thunderbirds* and *Stingray*, some of which were featured as KTMA-season episodes.

- Jack Perkins (Michael J. Nelson) - in real life the host of the A&E Network's *Biography* program, Perkins first appeared in *MST3K* simply to annoy the Mads by describing the movie with glowing praise. When

MST3K appeared in syndication as *The Mystery Science Hour*, Nelson's fake "Jack Perkins" hosted the show.

- Torgo (Michael J. Nelson) - a monster/henchman (supposed to be a satyr) in *Manos: The Hands of Fate*, Torgo was among the most frequently returning "guest characters" of *MST3K*. He got his knees fixed and returned as "Torgo the White" (an obvious parody of Gandalf the White) to accompany TV's Frank to "Second Banana Heaven" and was never seen again (episode 624, "Samson vs. the Vampire Women").

- Jan-in-the-Pan (Mary Jo Pehl) - a woman's severed head from the movie *The Brain That Wouldn't Die*.

- Pitch (Paul Chaplin) - a devil from the Mexican movie *Santa Claus*, Pitch was one of the few characters from the Comedy Central seasons to return in the Sci Fi Channel seasons.

- Santa Claus (Kevin Murphy) - appeared twice on the show, including a fight with Pitch, bellowing, "I'm here to chew candy canes and kick ass, and I'm all out of candy canes!"

- The Nanites (voiced variously by Kevin Murphy, Paul Chaplin, Mary Jo Pehl, and Bridget Jones) - self-replicating, bio-engineered organisms that work on the ship, they are microscopic creatures that reside in the S.O.L.'s computer systems. (They are similar to the creatures in *Star Trek: The Next Generation* episode "Evolution", which featured "nanites" taking over the *Enterprise*.) The Nanites made their first appearance in season 8. Based on the concept of nanotechnology, their comical *deus ex machina* activities included such diverse tasks as instant repair and construction, hairstyling, performing a Nanite variation of a flea circus, conducting a microscopic war, and even destroying the Observers' planet after a dangerously vague request from Mike to "take care of [a] little problem". They also ran a microbrewery. The Nanites were largely forgotten about during the show's last season, and we are not given an explanation of their fate following the series finale.

- Kevin Murphy played Wade the Nanite.
- Paul Chaplin played Ned the Nanite.
- Mary Jo Pehl played Jody the Nanite and Shelli the Nanite.
- Bridget Jones played Slicer the Nanite.

- Ortega (Paul Chaplin) - an unintelligible, decrepit, cigar-smoking henchman from the movie *The Incredibly Strange Creatures Who Stopped Living and Became Mixed-Up Zombies*, Ortega recurred occasionally in the three Sci Fi seasons.

- "Krankor" (Bill Corbett), who appeared in a host segment during the "Prince of Space" episode, and returned three episodes later in a host segment for "Invasion of the Neptune Men", featuring a movie with a similar plot. He was a vain, would-be conqueror with an unfortunately chicken-like appearance and a drawn-out, braying laugh (described by BBI as "like a Buick not turning over").

13.3 References

Chapter 14

List of Mystery Science Theater 3000 episodes

This page is a list of episodes for the American TV series *Mystery Science Theater 3000*. So far, there are a total of **197** television episodes and a feature film.

An 11th season of 14 new episodes is expected to be released later this year.

14.1 Series overview

14.2 Explanation of entries

Each entry starts with a code, which is its production number. Episodes are listed in production number order, not production order or air date order. The *MST3K* episode title is next, and if the original film title is different from the MST3K episode title, the former follows in parentheses. In the third line is the film's initial release year, a color/black & white notation, the production company (if known), and the country of origin. The last field shows the initial airdate of the episode (YYYY-MM-DD).

14.3 KTMA-TV

14.4 The Comedy Channel/Comedy Central

14.5 *Mystery Science Theater 3000: The Movie*

14.6 Sci Fi Channel

14.7 See also

- Elvira's Movie Macabre

- Mystery Science Theater 3000 video releases

- RiffTrax

14.8 References

14.9 External links

- Episode list for "Mystery Science Theater 3000" on Internet Movie Database

- List of episodes at TV.com

- Full episode guide at TV.com

Chapter 15

List of RiffTrax

The following is a list of RiffTrax, downloadable audio commentaries featuring comedian Michael J. Nelson heckling (or riffing on) films in the style of *Mystery Science Theater 3000*, a TV show in which Nelson was the head writer and, later, host. The RiffTrax are sold online where users can purchase and download the commentaries. The site was launched by Nelson and Legend Films in 2006 and is based in San Diego.

15.1 Official RiffTrax[1]

15.1.1 Commentaries[2]

The following is a list of films for which Michael J. Nelson and guest riffers have provided audio commentary, in order of the release date, for RiffTrax.

15.1.2 RiffTrax Presents

These are RiffTrax in which Michael J. Nelson does not appear.[3]

15.1.3 Public-domain shorts

These are presented as pre-synchronized video files, as opposed to audio files that users must synchronize themselves.[4]

15.1.4 Full-length VOD films

These are presented as pre-synchronized video files, as opposed to audio files that users must synchronize themselves.[4]

Note: The 7 original Mike-solo VODs were also released on dual-audio (Riffed and UnRiffed) DVD by Legend Films. *Missile to the Moon* was also released, but does *not* contain the Mike & Fred Willard riff.

15.1.5 RiffTrax DVDs and Blu-rays

These are RiffTrax featuring Mike, Kevin and Bill on a DVD or Blu-ray with the original film.[5]

Please note that *The Incredible 2-Headed Transplant* listed above is not an official release from RiffTrax. The RiffTrax audio track was licensed by Kino Lorber as a special feature for their DVD and Blu-ray release of the film.

15.1.6 Music[6]

15.1.7 Other downloads[7]

15.2 RiffTrax Live! events

The following is a list of Fathom Events's official Rifftrax Live! theatrical showings.

In October and November 2015, RiffTrax had a poll on their website asking users to choose two previous RiffTrax Live shows, up to and including *Miami Connection*, that they would like to see return to theaters in one night only encore presentations under the title "Best of RiffTrax Live!". The winners were *Starship Troopers*, rebroadcast on January 14, 2016 and *The Room*, rebroadcast on January 28, 2016.

15.3 References

[1] "Official RiffTrax". RiffTrax. Retrieved 2008-08-20.

[2] http://www.rifftrax.com/mp3-commentaries

[3] "RiffTrax Presents". RiffTrax. Retrieved 2008-08-20.

[4] "On Demand Catalogue". RiffTrax. Retrieved 2008-08-20.

[5] "Gift Shop :: DVDs". RiffTrax. Retrieved 2009-04-14.

[6] http://www.rifftrax.com/songs

[7] http://www.mst3kinfo.com/rifftrax/index.html

15.4 External links

- Official website

Chapter 16

Mike Nelson (character)

For the character "Mike Nelson" of *Sea Hunt*, see Sea Hunt.

Mike Nelson (full name Michael John Nelson)[1] is a fictional character in the comedy science fiction television series *Mystery Science Theater 3000*. Portrayed by actor/head writer Michael J. Nelson, Mike is a likable, sometimes dim temp worker from Wisconsin who comes to work for the mad scientists ("Mads") Dr. Clayton Forrester and TV's Frank in Deep 13 while they prepare for an evil-scientist audit in episode 512, *Mitchell.*

When Joel Robinson escapes from the Satellite of Love at the end of this episode, the "Mads" knock Mike unconscious and shoot him up to the satellite to replace Joel as their experimental guinea pig. The hapless Mike finds himself forced to watch bad movies with robot companions Tom Servo and Crow T. Robot while interjecting humorous quips and cultural riffs based on the action and dialog of the films. Nelson's first full appearance was episode 513, *The Brain That Wouldn't Die.* He typically wore either a dark green jumpsuit, a teal jumpsuit or a blue jumpsuit.

Series creator Joel Hodgson reportedly chose Nelson personally as his replacement, on the grounds that Nelson was a natural leader, a gifted comic and that he simply looked good standing next to the show's puppets. Prior to his tenure as host, Nelson played various parody "guest star" characters such as Torgo from Manos: The Hands of Fate, Morrissey, body builder and Hercules star Steve Reeves, and continued to play the occasional side character even after he became host. He also occasionally appeared as "Jack Perkins," and continued the persona as host of the *Mystery Science Theater Hour.*

16.1   Overview

During his tenure as the show's host, Nelson terminated the invention exchanges (last exchange during episode 519, Outlaw) and letter readings (episode 705, *Escape 2000*) that were a staple of the show's first five years, ending some popular traditions but giving the writers much more freedom in creating the opening and closing sketches. As co-star and writer Kevin Murphy explained, Nelson was many things "but he's not a tinkerer"; the invention exchanges had in any case been vehicles for Joel to engage in the sort of prop-based comedy he had specialised in before MST3K, and Nelson's strengths lay more in the portrayal of comic characters. Unlike Joel, who was resigned to his fate at being stranded in space, Mike is more desperate and proactive in trying to escape, making wild attempts during his run of the show. At one point, after seeing a bad Joe Don Baker film, he assumed he would leave as Joel did and boxed himself in a case of Hamdingers. However, what he thought was an escape pod was in fact the ship's water heater.

Mike's relationship with the bots was very different from the one the bots had with Joel. Though the bots occasionally riffed on Joel, they generally respected him and his authority, often regarding him as a father figure (unsurprising, since he literally created them) who would espouse life defining lessons that could neither be questioned nor refuted. The bots never felt that way about Mike. Though they came to accept him as their friend and companion, they were often more cynical and merciless with him (particularly during the Sci-Fi Channel era) when he screwed up or failed to pick up on something that was blatantly obvious to them, or on his history, such as dropping out of college (Mike had attended the University of Wisconsin-Stout). If Joel was a father figure, Mike was more like an older brother: while the bots were happy to make fun of him and play pranks on him, they were also willing to come to him when in need of comfort or advice. Mike was also different from Joel in personality and temperament. Joel tended to take his captivity in benign stride, often delivering his riffs in dead pan, holding no malice against his captors and affectionately calling them "the Mads". Mike tended to be more aggressive regarding his captivity and making fun of the films, and was forever scheming to escape the Satellite or at least make the Mads look like fools, though he and the bots did share moments of friendly recognition with later villains Pearl Forrester, Professor Bobo and Observer.

According to *The Mystery Science Theater 3000 Amazing Colossal Guide*, Mike is "as intelligent as the average man", and often displays a comic unawareness of his limitations and foibles. In episode #602 *Invasion U.S.A.*, he builds a robot whose only function appears to be to "destroy...destroy...destroy...". During the eighth season, he manages to contribute accidentally to the destruction of planets in three separate incidents. During *Mystery Science Theater 3000: The Movie*, one of Mike's mistakes was trying to show off by piloting the Satellite of Love, and ended up crashing into the Hubble Space Telescope before totally destroying it while attempting to get it off the ship. In episode #911 *Devil Fish*, he believes his identity has been stolen by a secret government agency merely after misplacing his wallet. He has an older brother, Eddie (also played by Nelson), a chain-smoking, hard-drinking bum who replaces him temporarily in the *Time Chasers* episode due to Crow going back in time to prevent him from getting the temp jobs which eventually result in him ending up on the SOL. However, this also results in Mike dying a horrible death, which caused Crow to go back once more to fix things. Mike is also tapped by Mrs. Forrester to provide "a little distraction" while she, Bobo and Brain Guy evacuate. This leads to Mike building a bomb out of baking soda and vinegar, and while preparing the bomb, he reminisces other incidents in which he built such bombs (as "harmless pranks") and ends up lost in traumatic memories, overloading the bomb with baking soda. The blast subsequently blows up the planet, causing Crow to ask, "Are you out of your stupid, rotted skull, you dumb man?!"

In the final episode of the series, Mike and the bots finally return to Earth, not by means of cunning escape but by the stupidity of their captors who accidentally send the Satellite tumbling back into Earth's atmosphere. The final scenes show Mike and his robot friends sometime later, now living together in a ground level apartment in Wisconsin, where Mike enjoys a bowl of rice as he sits down to watch *The Crawling Eye*, which had been the featured movie for the very first nationally syndicated episode of MST3K.

16.2 Notes

[1] Episode 815 - Agent for H.A.R.M

Chapter 17

MSTing

MSTing /ˌɛmˌɛsˈtiː.ɪŋ/ or **MiSTing** /ˈmɪstɪŋ/ is a method of mocking a show in the style of the television series *Mystery Science Theater 3000* (*MST3K*)[1] and, in particular, is a form of fan fiction in which writers mock other works by inserting humorous comments, called "riffs", into the flow of dialogue and events.[2]

17.1 Style

In MSTing, the author picks a badly written piece of text—usually a Usenet post, web page, or some other source such as a rant, spam or fan fiction—and inserts mocking comments from fictional readers of the text, essentially writing a script as if the MSTing were a movie. While "standard" MSTings attribute these comments to the three main characters of the *MST3K* cast, others might use characters - usually (though not always) from the universe of the story being mocked.[3] Often a prologue, epilogue, and intermissions are added in which the characters discuss a topic on the same theme as the original text, although intermission segments are usually dropped if the original work is short. Over time, the term MST has also been used to describe general fan fiction mockeries, without the use of the *MST3K* character-based joke format.

17.2 History

MSTing began in the early 1990s, as fans of the show, many of whom were involved in Usenet discussions in groups such as popular *MST3K* newsgroup *rec.arts.tv.mst3k.misc*, began adding amusing or critical remarks to others' posts, attributing them to the show's characters (typically, Crow T. Robot, Tom Servo, Joel Robinson, and later Mike Nelson).

The MSTing Mine credits Eric Alfred Burns as the writer of the first MSTings in February 1993, three short lambastings of Internet criticisms of *MST3K* itself.[4] The "canonical" MSTing style mostly derives from these first three posts, including Usenet-style quoting of the original work and script-style lines for the riffers and characters outside the theater. Other authors followed suit with MSTings of other works. By the end of the year, at least 29 original MSTings had been posted.[4]

As the phenomenon grew, it spread to other media and other forums. The newsgroup *alt.tv.mst3k.mstings* was established on Usenet for this fan writing.[5] A discussion category on Yahoo!, "Entertainment > Humor, Jokes, and Fun > Parody > Usenet Parodies > MiSTings" (no longer active in 2007), was created for discussions there.[6] MSTing for popular TV shows such as *Star Trek: Voyager*,[7] other genres such as anime,[8] Usenet postings, [9] and the MSTiers' own original works[10] were fodder for this written mockery. Recently, technologies not present when MSTing was started, such as Wikis and YouTube have been used.[11]

17.3 Issues and ethics

Generally for copyright and ethical reasons, "MSTers" attempt to gain the permission of the original story's writer before writing an MST treatment of the work in question. While it is not clear that this is legally or ethically necessary—the Fair Use doctrine technically permits reproduction for the purpose of parody—such permission is usually sought out of courtesy. As with the television version, some authors of works that have been made into "MSTings" have been rather negative about the treatment of their works and have requested the "MSTing" to be removed (which is generally done when requested), while others appreciate the humor and light criticism the collective result brings. Works of some of these latter authors (notably Stephen Ratliff for his *Star Trek: The Next Generation* fanfiction, in which teenage character Marissa Picard became an ensign in Starfleet shortly after her 14th birthday) became works that many "MSTers" seek to use as a basis for a "MSTing".

Additionally, some fiction sites such as FanFiction.Net have banned the posting of MSTs, either because they do not allow "script format" works (i.e. formats that are not standard for short stories, novellas, or novels, including those in teleplay or screenplay format, or those meant to imitate chat room conversations), or because they contain copyrighted material not written or owned by the MST's creator, or because the sites in question simply do not want to deal with upset authors getting angry about MSTings of their work.

17.4 References

[1] "Sequential Tart's Guide to Anime, Manga, and Miscellaneous Japanese Terms: A Glossary". *Sequential Tart - A Comics Industry Web Zine*. Retrieved 2007-01-30.

[2] "A.R.Yngve parodierar sin egen roman". *Mitrania*. Retrieved 2007-01-30.

[3] "Ms. Nitpicker's Fanfic Glossary". Retrieved 2007-01-30.

[4] "Index of old MSTings". *The MSTing Mine*. Retrieved 2007-01-27.

[5] "MST3K FAQs: MSTie Cyberspace". *The Satellite News*. Best Brains. Retrieved 2007-01-27.

[6] "Web Site Spotlight: March, 1998". *The Satellite News*. Best Brains. Archived from the original on 2007-09-27. Retrieved 2007-01-27.

[7] Spencer, S. (16 September 2005). "Zorak's Voyager MSTing Archive". Retrieved 2007-01-27.

[8] McLees, Tim. "Shuuichi's Vault of Anime Mistings". Retrieved 2007-01-30.

[9] Mamer, Karl (6 October 1996). "The Net: A Magnet For Bad Fiction". *Toronto Sun*. p. C15.

[10] "It Came from English 101!". River City Random. 13 November 2004. Archived from the original on 27 October 2009. Retrieved 2007-01-27.

[11] "WikiMsting". 25 January 2007. Retrieved 2007-01-30.

17.5 External links

- The MSTing Mine - An archive of MSTings originally posted to the MST3K Usenet newsgroups.

- The MSTers' Reference Centre

- EWIC MSTing FAQ Sheet

- How to Write MiSTings

- The Mystery Science Theater 3000 Fanvid and Live-performance Database

- A MSTing For All Seasons - Still active though sporadically updated.

- Atomic Monsters.com - Reviews of the cheesiest monster movies of the 1950s

- Project A.F.T.E.R. - Features fan fiction mockeries that do not use traditional character-based jokes.

- Fan-Friction - One of the only still-active MSTing webpages, featuring fan fiction and creepypasta mockeries.

- Court-Record's Sporking Theatre Where MSTings are called "sporks." Featuring fan fiction MSTings exclusively from the Ace Attorney fandom. Still active.

Chapter 18

Mystery Science Theater 3000

Mystery Science Theater 3000, abbreviated *MST3K,* is an American television comedy series created by Joel Hodgson and produced by Best Brains, Inc. The show premiered on KTMA in Minneapolis, Minnesota on November 24, 1988. It later aired on The Comedy Channel/Comedy Central for another six seasons until its cancellation in 1997. Through a fan-driven write-in campaign, the show was picked up by The Sci-Fi Channel and aired for another three seasons until another cancellation in August 1999. As of 2016, a new season is being produced and the platform it will be seen on is still unknown. The series ran for eleven years before a 15-year hiatus that ended in 2016, with 197 episodes and a feature film. The show also spun off *The Mystery Science Theater Hour*, a sixty-episode syndication package created by dividing selected episodes into halves.

The show initially starred Hodgson as Joel Robinson, a janitor trapped against his will by two mad scientists on the *Satellite of Love* and forced to watch a series of B movies as a part of the scientists' plot to take over the world. To keep his sanity, Joel crafts a number of robot companions — including Tom Servo, Crow T. Robot, and Gypsy — to keep him company and help him humorously comment on each movie as it plays, a process known as riffing. Each two-hour episode would feature a single movie in its entirety — with Joel, Tom, and Crow watching in silhouette from a row of theater seats at the bottom of the screen — and also included interstitial sketches. The show's cast changed over its duration; most notably, the character of Joel was replaced by Mike Nelson (played by Michael J. Nelson) in the show's fifth season. Other cast members, most of whom were also writers for the show, include Trace Beaulieu, Josh Weinstein, Jim Mallon, Kevin Murphy, Frank Conniff, Mary Jo Pehl, Bill Corbett, Paul Chaplin, and Bridget Jones Nelson.

MST3K was listed as one of *Time* magazine's "100 Best TV Shows of All" in 2007. The show won a Peabody Award in 1993, was also nominated for two Emmy Awards in the category of Outstanding Individual Achievement in Writing for a Variety or Music Program in 1994 and 1995, and was nominated from 1992 to 1997 for a CableACE Award. The show was considered highly influential, and partially con-tributing towards the practice of social television, while also bringing to light movies that had not received public attention and subsequently identified as some of the worst movies ever made, most notably *Manos: The Hands of Fate*. A large fanbase for the show grew during the onset of Internet growth in the 1990s, with fans calling themselves "MSTies".

Following the show's cancellation, various crew members launched separate projects in the same theme as *MST3K*, including *RiffTrax* (which continues to be ongoing) and *Cinematic Titanic*. A plan to revive the series was launched in 2015 by Hodgson and Shout! Factory, who has helped to secure licensing rights for the *MST3K* brand and for past *MST3K* episodes for home media and online streaming. The revival, funded through a successful crowd-sourced Kickstarter campaign, is expected to include fourteen new episodes, featuring Jonah Ray as the new human test subject aboard the *Satellite of Love*, with Felicia Day and Patton Oswalt as the new Mads. The revived series is anticipated to be released in the latter part of 2016.

18.1 Premise

MST3K is set in the "not-too-distant future." Two mad scientists, Dr. Clayton Forrester (Trace Beaulieu) and his sidekick Dr. Laurence Erhardt (Josh Weinstein), launch Joel Robinson (Joel Hodgson), a janitor working for Gizmonic Institute, into space aboard the orbiting dogbone-shaped *Satellite of Love*. Forrester and Erhardt — collectively referred as "The Mads" on the show — operate the *Satellite of Love* from their secret Deep 13 underground base, and force Joel to watch a series of B-movies in order to pinpoint the perfect B-movie to use as a weapon in Dr. Forrester's scheme of world domination.

To keep his sanity, Joel builds several sentient robots collectively named "the 'bots": Tom Servo; Crow T. Robot; Gypsy, who is in charge of running the satellite's operations; Cambot, the silent recorder of the experiments; Magic Voice, a disembodied female voice offering various announcements during segments of the show; and Rocket Number Nine, an camera-bot external to the *Satellite*. Joel has no control over when the movies start, because he used the parts that would have allowed him to do so to build the robots. He must enter the theater when the movie is sent up, because the Mads have numerous ways to punish Joel for non-compliance, including shutting off the oxygen supply to the rest of the ship and electric shocks. As the movie plays, Joel, Tom Servo, and Crow wisecrack and mock the movie — a practice they often referred to as "riffing" — to prevent themselves from going mad.

Over the course of the show's run, there were several cast changes, with the show's narrative often adjusted to match. When Weinstein left the series after the first national season, Kevin Murphy replaced him as the voice of Tom Servo while TV's Frank (Frank Conniff) replaced Weinstein's Dr. Erhardt as Dr. Forrester's lackey. Hodgson departed the series halfway through the fifth season; head writer Michael J. Nelson (playing a new character, Mike Nelson) replaced him as the show's human host until the end of the series. When Conniff left following the sixth season, Dr. Forrester was paired with his mother Pearl Forrester (Mary Jo Pehl) for the seventh season. Bealieu left MST3K following the seventh season; when the show returned on the Sci-Fi Channel, Bill Corbett took over as Crow, while Pearl Forrester was promoted to lead "Mad", aided by the alien Observer (Corbett) and the anthropomorphic ape Professor Bobo (Murphy).

18.2 Format

Episodes of *Mystery Science Theater 3000* are generally 90 minutes in running time, or 2 hours with broadcast advertisement breaks. Each episode primarily features the riffing of the movie, with these theater segmented wrapped with live-action skits performed by the cast. The introductory sketch is typically unrelated to the remainder of the episode, and followed by an introduction to the movie by the Mads. During Hodgson's period on the show, the introductory skits would typically involve an "Invention Exchange," where Joel would present a new invention to the Mads, and vice versa. This was an extension of Hodgson's own prop comedy aspects, and while they were continued into the Nelson era, they were ultimately dropped as, according to Murphy in *The Amazing Colossal Episode Guide*, "Joel was the gizmocrat, the one who brought that invention exchange spirit on board," while "Mike is many things, but he is not a tinkerer." The introductory segments would end with lights flashing and sirens blaring on the bridge of the *Satellite of Love*, and the crew running around in a panic and announcing that "We've got movie sign!" The scene would transition from the bridge to the theater on the op-

posite side of the *Satellite* via a "door sequence," where the camera would pass through six doors before the theater was revealed; similarly, the reverse of this shot was used to transition from the theater back to the bridge.

In the theater, Joel or Mike, Crow, and Tom would sit in silhouette in a row of theater seats and watch the movie, often with Joel or Mike using their hands to point and mock the movie in addition to their verbal riffing. Infrequently, the silhouette format was used for jokes, including as a means of unobtrusive censor bars for certain films. In many episodes featuring movies too short to fill the show's running time, the movie would be preceded by one or more shorts, educational films, newsreels, or similar material in the public domain. In other cases, longer movies were trimmed to fit the running time.

Interstitial skits would be used around commercial breaks, and a final skit ended the show. Skits would often, but not always, be related to the shorts or movies being shown. Many skits would feature "guest characters" (often from or inspired by the movie being featured, or from a past featured movie), often by way of the *Satellite of Love*'s "Hexfield Viewscreen" or through Rocket Number Nine. While these were generally played by the Best Brains crew in makeup (such as Nelson as Torgo from *Manos: The Hands of Fate*), both Minnesota Vikings running back Robert Smith and film critic Leonard Maltin have appeared as guests.

The final skit in Hodgson's period usually included reading fan mail and advertising the MST3K Info Club. This was phased out near the end of the Comedy Central run for the show. Shows with Dr. Forrester and TV's Frank would nearly always end with Dr. Forrester telling Frank to "push the button" to terminate the transmission. Almost all shows feature a stinger following the end credits of the show, typically a short humorous clip taken out of context from the film.

A limited selection of episodes were redeveloped into an hour-long *Mystery Science Theater Hour*, which enabled Best Brains to offer the show in syndication. In these, the episode was split into two parts, with new skits leading and ending each hour of Nelson portraying television host Jack Perkins in a parody of Perkins' *Biography* series.[3]

18.3 Production

18.3.1 Influences

Prior to *MST3K*'s 1988 debut, the nationally syndicated TV series *Mad Movies with the L.A. Connection* and *The Canned Film Festival* similarly made fun of many of the same movies. Each show lasted a single season, in 1985 and 1986, respectively.

Hodgson is credited for devising the show's concept. Prior to the show, Hodgson was an upcoming comedian having moved to Los Angeles and made appearances on *Late Night with David Letterman* and *Saturday Night Live*. He had been invited by Brandon Tartikoff to be on a NBC sit-com co-starring Michael J. Fox, but Hodgson felt the material was not funny and declined.[4] He further became dissatisfied with the Hollywood attitudes when they tried to double their offer, earning what he called a "healthy disrespect" of the industry.[5] He moved back to Minneapolis-St Paul, taking a job in a T-shirt printing factory that allowed him to conceive of new comedy ideas while he was bored. One such idea was the basis of *MST3K*, a show to riff on movies and that would also allow him to showcase his own prop comedy-style humor.[6]

The illustration for the song "I've Seen That Movie Too" in the liner notes of Elton John's Goodbye Yellow Brick Road, *which Hodgson took inspiration for* MST3K*'s theme and approach*

Hodgson said that part of the idea for *MST3K* came from the illustration for the song "I've Seen That Movie Too" (drawn by Mike Ross) in the liner notes from Elton John's *Goodbye Yellow Brick Road* album, showing silhouettes of two people in a theater watching a movie.[6] Hodgson also likened the show's setting to the idea of a pirate radio station broadcasting from space.[7] Hodgson credits *Silent Running*, a 1972 science-fiction film directed by Douglas Trumbull, as being perhaps the biggest direct influence on the show's concept. The film is set in the future and centers on a human, Freeman Lowell (Bruce Dern), who is the last crew member of a spaceship containing Earth's last surviving forests. His remaining companions consist only of three robot drones. *MST3K* and the Joel Robinson character occasionally reflected Lowell's hippie-like nature.[6][8] Hodgson wanted the feel of the show to appear homemade, and

cited the example of a crude mountain prop used during the *Saturday Night Live* sketch "Night on Freak Mountain" that received a humorous reaction from the studio audience as the type of aesthetic he wanted for the show.[8]

Both old movies and music inspired several of the show's character names as developed by Hodgson. The show's name came from the promotional phrase "Mystery Scientist" used by magician Harlan Tarbell and a play on the name of Sun Ra's band, the Myth Science Arkestra.[8] The "3000" was added to spoof the common practice of adding "2000" to show and product names in light of the upcoming 21st century, and Hodgson thought it would set his show apart to make it "3000".[8] Dr. Forrester was named after the main character of *The War of the Worlds*. The *Satellite of Love* was named after the song of the same name by Lou Reed.[7] Crow T. Robot was inspired by the song "Crow" from Jim Carroll's *Catholic Boy*,[7] while Rocket Number 9's name was inspired by the original name of Sun Ra's album *Interstellar Low Ways*.[7]

18.3.2   Writing

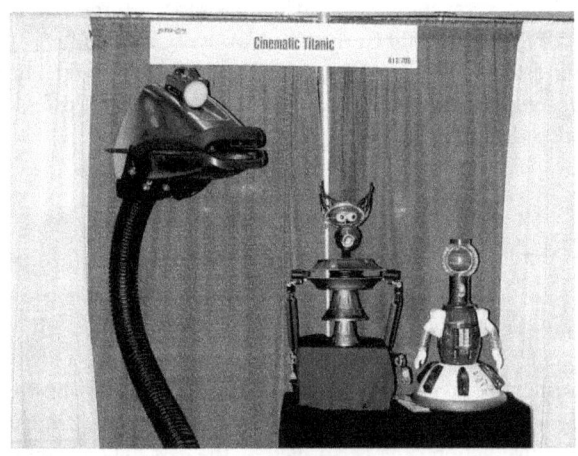

The 'bots of MST3k *as they appeared through the majority of its run: Gypsy (left), Crow T. Robot, and Tom Servo. The 'bots were created by Hodgson and fashioned out of common household objects.*

In the initial KTMA days, Mallon would present the writers with selections of movies from the station's archive to work from. In subsequent seasons, movie options were provided by the cable network.[6] To assure that they would be able to produce a funny episode, at least one member of the staff would watch the suggested films completely, generally assuring that the movie would be prime for jokes throughout; Conniff stated that he often would have to watch twenty films in their entirety before selecting one to use for the show.[9] In one specific case, the second season episode with the film *Sidehackers*, they had only skimmed the first part of the movie before making the decision to use it, and only later discovering that it contained a rape scene. They decided to stay committed to the film, but cut out the offending scene and had to explain the sudden absence of the affected character to the audience.[9] Since that point, they carefully scrutinized entire films, and once one was selected and assured the rights, committed to completing the episode with that film.[6] Obtaining the rights was handled by the cable networks. Some licensing required buying film rights in packages, with the selected bad movies included in a catalog of otherwise good films, making the negotiations odd since the network was only interested in the bad film. Other times, the actual rights to the film were poorly documented, and the network would follow the chain of custody to locate the copyright owner as to secure broadcast rights.[6]

During the KTMA era, the riffs during the movies were ad-libbed after making preliminary notes on the film's contents. In subsequent seasons, riffs were scripted by the writers.[6][10] An average episode (approximately 90 minutes running time) would contain more than 600 such riffs,[10] and some with upwards of 800 riffs.[11] Riffs were developed with the entire writing staff watching the film together several times through, giving off-the-cuff quips and jokes as the film went along, or identifying where additional material would be helpful for the comedy. The best jokes were polished into the script for the show.[6] Riffs were developed to keep in line with the characterization of Joel, Mike, and the 'bots.[6] Further, the writers tried to maintain respect for the films and avoided making negative riffs about them, taking into consideration that Joel, Mike, and the 'bots were companions to the audience while watching the movie, and they did not want to come off sounding like jerks even if the negative riff would be funny.[6][12] Hodgson stated that their goal in writing riffs is not to ridicule films as some often mistaken, and instead consider what they are doing as "a variety show built on the back of a movie".[13]

18.3.3 Filming

Production of an average episode of *MST3K* took about five to nine days once the movie was selected and its rights secured.[11][14] The first few days were generally used for watching the film and scripting out the riffs and live action segments. The subsequent days were then generally used to start construction of any props or sets that would be needed for the live action segments while the writers honed the script. A full dress rehearsal would then be held, making sure the segments and props worked and fine tuning the script. The host segments would then be filmed on one day, and the theater segments on the next. A final day was used to review the completed work and correct any major flaws they caught before considering the episode complete.[14] Live scenes used only practical special effects, and there

An example of MST3K*'s "Shadowrama" effect used as the central motif for the show. Here, Tom Servo (left), Joel Robinson, and Crow T. Robot, in silhouette, are watching the short* Mr. B Natural *in the 1991 episode featuring* War of the Colossal Beast

was minimal post-editing once filming was completed.[15]

The theater shots, the primary component of an episode, is filmed in "Shadowrama", a term trademarked by Best Brains; this appears as a row of theater seats with silhouettes of Joel or Mike, Crow, and Tom to one side, appearing to watch the movie on a big theater screen. In reality, the "seats" are a black-painted foamcore board sitting behind the seat (towards the camera) for Joel or Mike, and stages for the Crow and Tom puppets. The human host wore black clothing while the robot puppets were painted black; the screen they watched was a white luma key screen as to create the appearance of silhouettes. The actors would follow the movie and the script through television monitors located in front of them, as to create the overall theater illusion.[16]

To transition from skit segments to the theater segments, they created the "door sequence", which Hodgson took inspiration from the *Mickey Mouse Club*, noting that the commonality to the title credits of *Get Smart* were coincidental.[8] In devising this sequence, this also led to Beaulieu creating the dogbone-like shape of the *Satellite of Love* with additional inspiration taken from the bone-to-ship transition in the film *2001: A Space Odyssey*.[8] Hodgson had wanted to use a "motivated camera" for filming, a concept related to motivated lighting; in this mode, all the shots would appear to have been taken from an actual camera that was part of the scene to make the scene appear more realistic. This led to the creation of Cambot as a robot that Joel or Mike would speak to during host segments or filming them while in the theater, and Rocket Number Nine to show footage outside of the Satellite of Love.[17]

The show's theme song, the "Love Theme from Mystery Science Theater 3000", was written by Hodgson and Weinstein, which helped to cement some of the broader narrative elements of the show, such as the Mads and Joel being part of an experiment.[7] The song was composed by Charlie Erickson with help from Hodgson in the style of Devo, The Replacements, and The Rivieras (particularly their cover of the song "California Sun") and sung by Hodgson.[7][8] Initial shows used foam letters to make the show's title, but they later created the spinning-moon logo out of a 2 feet (0.61 m) diameter fiberglass ball, covered with foam insulation and the lettering cut from additional foam pieces. Hodgson felt they needed a filmed logo with the rotating effect as opposed to a flat 2D image, and though they had envisioned a more detailed prop, with the letters being the tops of buildings on this moon, they had no time or budget for a project of that complexity and went with what they had available.[18] Musical numbers would also be used as part of the host segments, which Hodgson said came out naturally from the riffing process; they would find themselves at times singing along with the movie instead of just riffing at it, and took that to extend songs into the host segments.[7]

18.4   History

18.4.1   KTMA era (1988–1989)

Hodgson approached Jim Mallon, at the time the production manager of KTMA, a low-budget local television station, with his idea of a show based on riffing on movies, using robots that were created out of common objects.[6] Mallon agreed to help produce a pilot episode, and Hodgson hired on local area comedians J. Elvis Weinstein (initially going by Josh Weinstein but later changed to J. Elvis as to distinguish himself from Josh Weinstein, a well-known writer for *The Simpsons*)[10] and Trace Beaulieu to develop the pilot show.[6] By September 1988, Hodgson, Mallon, Weinstein, and Beaulieu shot a 30-minute pilot episode, using segments from the 1968 science-fiction film *The Green Slime*.[6] The robots and the set were built by Hodgson in an all-nighter. Joel watched the movie by himself, and was aided during the host segments by his robots, Crow (Beaulieu), Beeper, and Gypsy (Weinstein). Hodgson used the narrative that his character named "Joel Hodgson" (not yet using his character name of Robinson) had built the *Satellite of Love* and launched himself into space.[19] Camera work was by Kevin Murphy, who worked at television station KTMA. Murphy also created the first "doorway sequence" and theater seat design. These initial episodes were recorded at the since-defunct Paragon Cable studios and customer service center in Hopkins, Minnesota.

Mallon met with KTMA station manager Donald O'Conner the next month and managed to get signed up for thirteen episodes. Show production was generally done on a 24-hour cycle, starting with Mallon offering a few films from KTMA's library for the writers to select from.[6] The show had some slight alterations — the set was lit differently, the robots (now Crow, Servo and Gypsy) joined Joel in the theater, and a new doorway countdown sequence between the host and theater segments was shot. The puppeteers worked personalities into their robots: Crow (Beaulieu) was considered a robotic Groucho Marx, Tom Servo (Weinstein) as a "smarmy AM radio DJ", and Gypsy (Mallon) modeled after Mallon's mother had a "heart of gold" but would become disoriented when confronted with a difficult task.[6] The development of the show's theme song would lead to establishing elements for the show's ongoing premise.[8]

Mystery Science Theater 3000 premiered at 6:00 p.m. on Thanksgiving Day, November 24, 1988 with its first episode, *Invaders from the Deep*, followed by a second episode, *Revenge of the Mysterons from Mars* at 8:00 p.m. Initially, the show's response was unknown, until Mallon set up a phone line for viewers to call in.[6] Response was so great that the initial run of 13 episodes was extended to 21, with the show running to May 1989. Hodgson and Mallon negotiated to secure the rights for the show for themselves, creating Best Brains, Inc., agreeing to split ownership of the idea equally.[6] During this time a fan club was set up and the show held its first live show at Scott Hansen's Comedy Gallery in Minneapolis, to a crowd of over 600. Despite the show's success, the station's overall declining fortunes forced it to file for bankruptcy reoganization in July 1989. The station sold Mystery Science Theater 3000 to cable network, The Comedy Channel (now Comedy Central) that year.[20]

18.4.2   Comedy Central era (1989–1996)

Just as *MST3K*'s run at KTMA was ending, HBO, looking to build a stable of shows for their new Comedy Channel cable network, approached Best Brains and requested a sample of their material.[6] Hodgson and Mallon provided a seven-minute demo reel, which led to the network greenlighting *MST3K* as one of the first two shows picked up by the network. The network offered Best Brains a relatively small figure, $35,000 per episode, for the show, but allowed Best Brains to retain the show's rights.[5] Though the Comedy Channel executives wanted the show to be filmed in New York City, Best Brains insisted on keeping the production in Minnesota setting up an office and warehouse space in Eden Prairie for filming.[6][21] Hodgson stated it would have cost four times as much per episode to film in either New York or Los Angeles.[22]

With an expanded but still limited budget, they were able to hire more writers, including Mike Nelson, Mary Jo Pehl, and Frank Conniff, and build more expansive sets and robot puppets.[6] They created the characters of Dr. Forrester

MST3K *cast and crew Pehl (left), Beaulieu, Hodgson, Weinstein, and Conniff, as part of the post-show project,* Cinematic Titanic *in 2011*

(Beaulieu) and Dr. Erhardt (Weinstein) and crafted the larger narrative of each episode being an "experiment" they test on Joel. The cable network was able to give them a larger library of films to select from with the network chasing down any rights negotiations that were needed. Instead of ad-lib riffs in the theater, each show was carefully scripted ahead of time, with Nelson serving as head writer.[10]

MST3K was considered Comedy Channel's signature program, generating positive press about the show despite the limited availability of the cable channel nationwide.[6] After the second season, The Comedy Channel and rival comedy cable network HA! merged to become Comedy Central. During this period, *MST3K* became the cable channel's signature series, expanding from 13 to 24 episodes a year. To take advantage of the show's status, Comedy Central ran "Turkey Day," a 30-hour marathon of *MST3K* episodes during Thanksgiving, 1991. This tradition would be continued through the rest of the Comedy Central era. Though the show did not draw large audience numbers compared to other programming on Comedy Central, such as reruns of *Saturday Night Live*, the dedicated fans and attention kept the show on the network.[5]

Weinstein left the show after the first Comedy Channel season and Murphy replaced him as the voice of Tom Servo, portraying the 'bot as a cultured individual.[6] The Dr. Erhardt character was replaced by Conniff as "TV's Frank." Despite the fact that Frank was a lackey and not a "mad scientist," he and Forrester were collectively referred to as "The Mads."

Hodgson decided to leave the series halfway through Season Five due to his dislike of being on-camera and his disagreements with producer Mallon for creative control of the program.[23][24] Hodgson also stated that Mallon's insistence to produce a feature film version of the show led to his departure, giving up his rights on the *MST3K* property to Mallon.[25] Hodgson later told an interviewer: "If I had the presence of mind to try and work it out, I would rather have stayed. 'Cause I didn't want to go, it just seemed like I needed to."[18] Though they held casting calls for a replacement for Hodgson on camera, the crew found that none of the potential actors really fit the role; instead, having reviewed a test run that Nelson had done with the 'bots, the crew agreed that having Nelson (who had already appeared in guest roles on the show) replace Hodgson would be the least jarring approach.[6] In his final episode, Joel was forced to sit through the Joe Don Baker movie *Mitchell* before escaping the *Satellite of Love* and returning to Earth with help from Gypsy and "Mike Nelson" (played by Nelson), a temp worker hired by Dr. Forrester to help prepare for an audit from the Fraternal Order of Mad Science. To replace Joel and continue his experiment, Dr. Forrester sent Mike up in his place. The replacement of Joel by Mike would lead to an oft-jokingly "Joel vs Mike flame war" in the *MST3K* fandom at the nascent period of Internet availability, debating endlessly about who was the better host in the same manner as "Kirk vs Picard" discussions in the *Star Trek* fandom.[26]

Mystery Science Theater 3000: The Movie was produced during the later half of the Comedy Central era and had a very limited theatrical release in 1996 through Universal Pictures. It featured Mike and the bots subjected to the film *This Island Earth* by Dr. Forrester. Though well received by critics and fans, the film was considered a flop due to its limited distribution.[25]

Conniff left the show after Season Six, looking to get into showwriting in Hollywood.[6][21] Within Season Seven, the show introduced Dr. Forrester's mother, Pearl (played by writer Mary Jo Pehl). By this time under new leadership of Doug Herzog, Comedy Central had started creating an identity for its network, which would lead to successful shows like *Politically Incorrect*, *The Daily Show* and *South Park*, leaving *MST3K* as an oddity on the network taking up limited program space. Herzog, though stating that *MST3K* "helped put the network on the map" and that its fans were "passionate", believed it was necessary to change things around due to the show's declining and lackluster ratings.[27][28] The network cancelled *MST3K* after its seventh season.[6] In the last show of the seventh season, *Laserblast*, Dr. Forrester detaches the SOL from Deep 13 after his funding runs out, casting the satellite adrift in space. Parodying 2001: A Space Odyssey, they reach the edge of the Universe and become entities of pure consciousness, as Forrester sees a monolith-like giant videotape, and then turns into a star-baby.

18.4.3   Sci-Fi Channel era (1997–1999)

Nelson, Corbett, and Murphy, the primary actors in the Sci-Fi channel area, as part of their Rifftrax *panel in 2009*

When Comedy Central dropped the show after a six-episode seventh season, the show staff continued to operate for as long as they still had finances to work with.[29] *MST3K*'s fan base staged a write-in campaign to keep the show alive.[20] This effort led the Sci-Fi Channel, a subsidiary of USA Networks to pick up the series. Rod Perth, the president of programming for USA Networks at that time, helped to bring the show to the Sci-Fi Channel, stating himself to be a huge fan of the show, and believing that "the sci-fi genre took itself too seriously and that this show was a great way of lightening up our own presentation".[29]

MST3K ran for three more seasons on the Sci-Fi Channel. At this point, Beaulieu opted to leave the show, feeling that anything creative that would be produced by Best Brains would belong to Mallon, and wanted to have more creative ownership himself.[6] To replace Dr. Forrester, Pearl Forrester was given two new sidekicks: the idiotic, *Planet of the Apes*-inspired Professor Bobo (played by Murphy) and the highly evolved, supposedly omniscient, yet equally idiotic Observer (a.k.a. "Brain Guy"), played by writer Corbett. In addition, Corbett took over Crow's voice and puppetry and Best Brains staffer Patrick Brantseg took over Gypsy in the middle of Season Eight.[30] In the overarcing narrative, Mike and the 'bots return to corporeal form, and then are chased down by Pearl in the *Widowmaker*, a modified VW Van, before they return to Earth, after which Pearl sends movies to the *Satellite of Love* from Castle Forrester. With this replacement, the series' entire original cast had been turned over.

During the Sci-Fi era, Best Brains found themselves more

limited by the network: the pool of available films was smaller and they were required to use science fiction films,[31] and the USA network executives managing the show wanted to see a story arc and had more demands on how the show should be produced.[6] Conflict between Best Brains and the network executives would eventually lead to the show's 2nd cancellation.[6] Peter Keepnews, writing for the *New York Times*, noted that the frequent cast changes, as well as the poorer selection of films that were more dull than bizarre in their execution, had caused the show to lose its original appeal.[32] Another campaign to save the show was mounted, including several *MST3K* fans taking contributions for a full-page ad in the trade publication *Daily Variety* magazine, but was unsuccessful.[33]

The season 10 finale, *Danger: Diabolik*, premiered on August 8, 1999, during which, in the show's narrative, Pearl Forrester sent the *Satellite of Love* out of orbit, with Mike and the 'bots escaping and taking up residence in an apartment near Minnesota.[33] A "lost" episode produced earlier in the season but delayed due to rights, *Merlin's Shop of Mystical Wonders*, was the final season 10 episode of *MST3K*, broadcast on September 12, 1999.[20] Reruns continued to air on the Sci Fi Channel for several years, ending with *The Screaming Skull* on January 31, 2004. The shows later moved to syndication.

18.4.4   Revival (2016–ongoing)

revival will feature Ray (top) aboard the Satellite of Love, Day (bottom left) as Kinga Forrester, and Oswalt as TV's Son of TV's Frank

Since around 2010, Hodgson had been trying to bring back *MST3K*, with efforts increasing since 2013. Hodgson began working closely with Shout! Factory, the production company that has published many of the original *MST3K* episodes including securing the rights to the films contained within, to seek ways to create a revival. Hodgson felt that timing was right for the reboot: previous shows have found crowd sourced funding from its fans for continuation, and with non-traditional outlets for broadcast, such as Netflix, there is a potential for a wider audience.[34] Hodgson also considered that fans still appreciated the show and the cast and crew from it even 25 years from its premiere, and felt that a fan-enabled effort would help lead to new *MST3K* episodes. Hodgson had already considered that a wholly new cast would be involved in any revival project, noting that the original series had had a full cast change by the end of its run.[6][35] One obstacle towards a revival was the rights to the *MST3K* property, which were still held by Mallon and Best Brains. Shout! Factory worked with Hodgson, Mallon, and Best Brains as to acquire full rights to the show around August 2015.[15][36]

With the rights secured, Hodgson and Shout! Factory launched a $2 million Kickstarter campaign in November 2015 to produce at least three new episodes, with potentially up to 12 if a $5.5 million funding goal could be met.[37] Hodgson estimated that each episode would cost about $250,000 to make, in addition to five-figure movie licensing rights, in contrast to $100,000 needed for the original series.[38] Though Hodgson stated he had talked to various networks and streaming providers, he felt it was necessary to start with a crowd-sourced means to revive the show as to let the fans decide what type of show they wanted, instead of something that would be dictated by a network and may miss the mark.[39] Hodgson believed that the Kickstarter was necessary for "convincing a 'conference table full of executives' to give our show a home".[40] Hodgson, in speaking to *Rolling Stone* following the Kickstarter, stated that they had just been in discussion with a network that showed renewed interest in carrying the show due to the success of the fan-backed campaign.[15]

The Kickstarter campaign reached its base funding for a three-episode revival within a week of its launch.[41] On the final day of the campaign, Hodgson and Shout! ran a streaming telethon which included appearances from the newly selected cast and crew, and various celebrities that supported the revival to help exceed the target funding levels for twelve episodes.[42] The campaign ended on December 11, 2015 with total funding of $5,764,229 from 48,270 backers, with an additional $600,000 in backer add-ons, which allowed Hodgson to plan two more additional episodes, including a Christmas episode, to bring the total season to fourteen episodes.[43][44] The Kickstarter became the largest one for Film & Video, surpassing the $5.70 million raised for the *Veronica Mars* film.[45]

During the campaign, Hodgson announced the cast members that he had approached before the Kickstarter and was then able to confirm their involvement with the Kickstarter's success. Comedian Jonah Ray will play Jonah Heston, the new host aboard the Satellite of Love, watching and riffing on the films. Hodgson had met Ray while recording an episode of *The Nerdist Podcast*, and felt he would be a good

fit.[41] The voices of Crow and Tom will be provided by comedians Hampton Yount and Baron Vaughn, respectively, both whom Ray recommended to Hodgson.[46] Felicia Day will play Kinga Forrester, Clayton Forrester's daughter and one of the new Mads in charge of the experiments; Hodgson had seen Day's performance in shows like *The Guild* and *Dr. Horrible's Sing-Along Blog*, and felt she matched his idea for the character.[46] Patton Oswalt will play Kinga's henchman, TV's Son of TV's Frank; Hodgson had initially considered bringing on Oswalt, a longtime friend and self-professed *MST3K* fan, as a special guest writer for an episode of the revived series, but decided after the Kickstarter had already begun that Oswalt would also be a good fit as an on-camera performer.[47] In addition, Oswalt will help write for the show, whom Hodgson believes can readily handle the comedic nature of an *MST3K* episode.[38] Hodgson himself will remain primarily off-camera as the executive producer for the remake.[38]

The new show will be written by a new staff selected by Hodgson, as well as from the new cast members and from backers of the Kickstarter campaign at specific reward levels. Former head writer for *The Daily Show With Jon Stewart* Elliott Kalan will be the lead writer for the revival.[48] Hodgson has also announced plans to have guest writers for certain episodes that include Dan Harmon, Justin Roiland, Rob Schrab, Nell Scovell, Ernie Cline, Pat Rothfuss, Paul & Storm, and Dana Gould. Additionally, Robert Lopez will compose original songs for the new episodes.[49] Set and prop designers will include Wayne White, Pendleton Ward, Rebecca and Steven Sugar, and Guy Davis, while live and practical special effects will be planned out by Adam Savage.[15][50] Among other staff will include returning *MST3K* crew, including: Charlie Erickson, who composed the original show's theme song and will compose the new show's theme and other musical arrangements; Beez McKeever, who worked on the original show's props and will design costumes and props for the new show; Crist Ballas will continue doing hair and makeup design; and Paul Chaplin, one of the show's original writers to help write the new shows.[51] Hodgson has stated that Jack Black, Joel McHale, Bill Hader, Neil Patrick Harris, Jerry Seinfeld, and Mark Hamill have expressed interest in appearing in cameos on the new shows.[52]

Hodgson also opened up to the show any of the existing cast members to make cameo appearances or aid in the creative process. However, Nelson, Weinstein, Beaulieu, Pehl, and Corbett stated that they had declined to be involved with the MST3K reboot;[53][54] Nelson felt that "The brand does not belong to me, and I make and have made (almost) zero dollars off it since it stopped production in 1999."[54][55] Murphy recognized that as a relaunch, "a whole new group of people are brought in, and the show goes off in a new direction, and I think that's great".[56] Conniff noted on his

Twitter that Shout! Factory would be "cutting [the former cast members] in, financially at least" on the profits from the series.[57]

Production for the new season began on January 4, 2016.[58] In an interview with *Paste* during the Kickstarter, Hodgson expects to be able to share clips at the 2016 San Diego Comic-Con International in July.[39] Hodgson is also exploring the possibility of having the show on broadcast television.[15] Hodgson is aiming to follow in the pattern of what made for fan-favorite episodes from the original series, borrowing equally from the Joel and Mike eras; he noted there were about thirty episodes that he and fans universally agreed were the show's best, and expects to use these as templates as the basis of the new show.[15] The film selection was narrowed down to about twenty movies as of February 2016, with the rights secured to about half of them, while Shout Factory is working to assure worldwide distribution rights for the others;[58] Hodgson noted that the films will be more recent than those used on the original series, with "maybe one" from the 1950s/1960s, but does not want to reveal what these films are until the episodes are broadcast as to have the biggest comedic effect on the audience.[38]

18.5   Cast

Main article: List of Mystery Science Theater 3000 characters

Notes

1. ^ "Joel Hodgson" during season 0; Simply "Joel" (no last name) during Season 1.

2. ^ Simply "Frank" during seasons 2 and 3.

3. ^ Guest appearance only.

18.6   Episodes

Main article: List of Mystery Science Theater 3000 episodes

By the conclusion of the Sci-Fi era, a total of 197 *MST3K* episodes, have been produced.[59] This does not include *The Green Slime* pilot episode, which was used to sell the concept to KTMA but otherwise was never broadcast.[60]

None of the KTMA episodes were rerun nationally or have been released onto home video due to rights issues. Some consider the first three KTMA episodes to be "missing episodes", as no fan copies are known to exist, but master copies of all these episodes still exist according to Mallon.[61] The credits in the first four seasons on Comedy Central included the phrase "Keep circulating the tapes" to encourage fans to share VHS tapings they made with others, despite the questionable copyright practice. Though the phrase was removed from the credits, the concept of "keep circulating the tapes" was held by the show's fans to continue to help introduce others to the show following its broadcast run.[59]

18.6.1   Turkey Day marathons

A common event in both the Comedy Central and Sci-Fi eras was a Turkey Day marathon that ran on or near the Thankgiving holiday. The marathon would show between six and twelve rebroadcasts of episodes, often with new interstitial material between the episodes from the cast and crew.[62]

In honor of the show's 25th Anniversary in 2013, Shout! Factory ran a streaming video "Turkey Day" on Thanksgiving as had similarly been done during the show's run on Comedy Central. Fans were able to select the six episodes they wanted to see and the event was hosted by Hodgson.[63] The success of this event led Hodgson and Shout! Factory to repeat the event the following year.[64] In the final segment, Joel was joined at the dinner table by Crow T. Robot and Tom Servo. In 2014, another marathon was held. This time Crow and Tom were full participants, voiced by their original voice actors, Trace Beaulieu and Josh Weinstein, respectively.

The streaming Turkey Day event was run again 2015, coinciding with the Kickstarter for the planned revival of the show.

18.6.2   Home video and digital releases

Main article: Mystery Science Theater 3000 video releases

Home video releases of *MST3K* are complicated by the licensing rights of the featured film and any shorts, and as such many of the nationally-televised episodes have not yet been released onto home video. Through the current distributor, Shout! Factory, over 100 of the movies have been cleared for home media distribution.[65]

Original home media releases were issued by Rhino Entertainment, initially starting with single disc releases before switching to semi-regular four-episode volume set. According to Hodgson, the people at Rhino that were involved in the distribution of *MST3K* eventually left Rhino and joined with Shout! Factory, helping to convince that publisher to acquire the rights from Rhino.[65] Since 2008, all releases *MST3K* have been through Shout! Factory, (including some reprints of the first Rhino volume set) and have typically been multi-episode volumes or themed packs.

In 2014, 80 episodes of the show were made available for purchase or rental on the video streaming site Vimeo.[66] Shout! Factory has uploaded some episodes to YouTube with annotations, as documented by *The Annotated MST* fansite, to explain some of the sources of the jokes in the riffs.[67] In February 2015, Shout! Factory launched its own streaming service, Shout! Factory TV, of which selected episodes of *MST3K* were included on the service.[68] Selected episodes were also made available on demand through *Rifftrax* starting in November 2015.[69]

18.7   Adaptations

18.7.1   Syndication

In 1993, the show's staff selected 30 episodes to split into 60 one-hour segments for *The Mystery Science Theater Hour*. The repackaged series' first-run airings of these half-shows ran from November 1993 to July 1994. Reruns continued through December 1994, and it was syndicated to local stations from September 1995 to September 1996, allowing stations to run the series in a one-hour slot, or the original two hour version.[70] *MST3K* returned to television for the first time in ten years in July 2014, when RetroTV began broadcasting the series on Saturday nights, with an encore on Sunday evenings.[71] The following year, they started showing on regular PBS affiliate networks.[72][73]

18.7.2   Feature film

Main article: Mystery Science Theater 3000: The Movie

In 1996, Universal Studios released *Mystery Science Theater 3000: The Movie*, a film adaptation in which Mike and the bots riffed *This Island Earth*. The film was released on DVD in the United States by Image Entertainment. Universal Pictures re-released the film on DVD on May 6, 2008, with a new anamorphic widescreen transfer, Dolby Digital 5.1 Surround Sound mix, and the film's original trailer.[74]

18.7.3   Book

In 1996, the book, *The Amazing Colossal Episode Guide* (written by some *MST3K* cast members), was released, which contained a synopsis for every episode from seasons

one through six, and even included some behind-the-scenes stories as well. In it, Kevin Murphy related two tales about celebrity reactions he encountered. In one, the cast went to a taping of Dennis Miller's eponymous show; when they were brought backstage to meet Miller, the comedian proceeded to criticize the *MST3K* cast for their choice of movie to mock in the then-recent episode "Space Travelers" (a re-branded version of the Oscar-winning film *Marooned*).[75] Murphy also discussed how he met Kurt Vonnegut, one of his literary heroes. When he had mentioned the show and its premise to Vonnegut, the author suggested that even people who work hard on bad films deserve some respect. Murphy then invited Vonnegut to dine with his group, which Vonnegut declined, claiming that he had other plans. When Murphy and friends ate later that night, he saw Vonnegut dining alone in the same restaurant, and remarked that he had been "faced...but *nicely* faced" by one of his literary heroes.[76]

18.7.4   Other appearances

In 1996, during promotion for *Mystery Science Theater 3000: The Movie*, Mike and the bots were interviewed in-character on MTV, and seen in silhouettes heckling footage from MTV News. Also that year, Joel Hodgson was a featured guest on Cartoon Network's *Space Ghost Coast to Coast*. In 1997, the E! network's *Talk Soup* show, starring John Henson, featured guest appearances from Mike, Crow, and Tom Servo.

In 1997, the videogame magazine *PlayStation Underground* (Volume 2, Number 1) included a Best Brains-produced *MST3K* short on one of their promotional discs. The video opened with a host segment of Mike and the Bots playing some PlayStation games, only to go into the theater to riff on some videos from the magazine's past. The feature is about seven minutes long. An Easter egg on the disc has some behind-the-scenes footage of Best Brains filming the sequences.[77] Also that year, a new online animated web series, referred to as "The Bots Are Back!", was produced by Jim Mallon. The series planned a weekly adventure featuring Crow, Tom Servo, and Gypsy, with Mallon reprising his role as Gypsy and Paul Chaplin as Crow. However, only a handful of episodes were released, and the series was abandoned due to budget issues. The internet response to the webisodes was largely negative.[78]

18.8   Reception

In 2004, the show was listed as #11 in a featured *TV Guide* article, "25 Top Cult Shows Ever!", and included a sidebar which read, "Mike Nelson, writer and star (replacing creator Joel Hodgson), recently addressed a college audience: "There was nobody over the age of 25. I had to ask, 'Where are you seeing this show?' I guess we have some sort of timeless quality."[79] Three years later, *TV Guide* rewrote the article, and bumped *MST3K* to #13.[80] In 2007, the show was listed as one of *Time* magazine's "100 Best TV Shows of All".[81] In 2012, the show was listed as #3 in *Entertainment Weekly*'s "25 Best Cult TV Shows from the Past 25 Years", with the comment that "*MST3K* taught us that snarky commentary can be way more entertaining than the actual media."[82]

18.8.1   Reactions by those parodied

The reactions of those parodied by *MST3K* has been mixed. Sandy Frank, who held the rights to several *Gamera* films parodied on the show, was "intensely displeased" by the mockery directed at him. (The crew once sang the "Sandy Frank Song", which said that Frank was "the source of all our pain", "thinks that people come from trees", Steven Spielberg "won't return his calls", and implied that he was too lazy to make his own films). Because of this, Frank reportedly refused to allow the shows to be rebroadcast once *MST3K*'s rights ran out.[83] However, this may in fact be a rumor, as other rumors indicate that the *Gamera* films distribution rights prices were increased beyond what BBI could afford as a result of the show's success. According to Shout Factory, the Japanese movie studio Kadokawa Pictures were so horrified with MST3K's treatment of five *Gamera* films that they refused to let Shout release the episodes on home video. Brian Ward (one of the members of Shout Factory) explained to fans on the forums of the official Shout Factory website that they tried their best to convince them, but the Japanese take their *Gamera* films very seriously and do not appreciate their being mocked. However, eventually Shout was able to clear the episodes for a special 2011 release due to the rights in North America shifting away from the Japanese to another, North American entity that had no such qualms.[84] In another post on the Shout Factory message boards, Ward explained that the Godzilla films faced the same obstacle as *Gamera*, and explained that unless the rights shifted the way the *Gamera* rights have, these films would remain unreleased.[85]

Kevin Murphy had once said that Joe Don Baker wanted to beat up the writers of the show for attacking him during *Mitchell*.[86][87] Murphy later said Baker likely meant it in a joking manner, although Mike Nelson said that he deliberately avoided encountering Baker while the two happened to be staying at the same hotel.[88]

Director Rick Sloane was shocked at his treatment at the conclusion of *Hobgoblins*, in which Sloane himself was mercilessly mocked over the film's end credits.[89] In a 2008

interview, however, Sloane clarified his comments, saying that "I laughed through the entire MST3K episode, until the very end. I wasn't expecting the humor to suddenly be at my own expense. I was mortified when they dragged out the cardboard cutout and pretended to do an interview with me. I was caught off guard. I had never seen them rip apart any other director before on the show." He also credits the success of the *MST3K* episode with inspiring him to make a sequel to *Hobgoblins*, released in 2009.[90]

Jeff Lieberman, director of *Squirm*, was also quite angry at the *MST3K* treatment of his film.[91]

Others have been more positive: Robert Fiveson and Myrl Schriebman, producers of *Parts: The Clonus Horror*, said they were "flattered" to see the film appear on *MST3K*.[92] Actor Miles O'Keeffe, the star of the film *Cave Dwellers*, called Best Brains and personally requested a copy of the *MST3K* treatment of the film,[88] saying he enjoyed their skewering of what he had considered to be a surreal experience; according to Hodgson, O'Keefee said his friends always heckled his performance in the film when it was on, and he appreciated the *MST3K* treatment.[22] In the form of an essay and E. E. Cummings-esque poem, Mike Nelson paid tribute to O'Keeffe with a humorous mix of adulation and fear.[93]

Actor Adam West, star of the 1960s *Batman* TV series, co-starred in *Zombie Nightmare*, another film *MST3K* mocked. West apparently held no grudges, as he hosted the 1994 "Turkey Day" marathon in which the episode featuring *Zombie Nightmare* had its broadcast premiere. Mamie van Doren (who appeared in episode 112, *Untamed Youth*, and episode 601, *Girls Town*), Robert Vaughn (star of episode 315, *Teenage Cave Man*, which he called the worst movie ever made) and Beverly Garland (who had appeared in many *MST3K*-featured Roger Corman films) also hosted at the marathon.

Rex Reason, star of *This Island Earth*, has also appeared at several *MST3K* events and credits *MST3K* with introducing the film to a new generation. The crew of *Time Chasers* held a party the night the *MST3K* treatment of their film aired and, while reactions were mixed, director David Giancola said, "Most of us were fans and knew what to expect and we roared with laughter and drank way too much. I had a blast, never laughed so hard in my life."[94]

18.8.2   Awards

In 1993, *MST3K* won a Peabody Award for "producing an ingenious eclectic series": "With references to everything from Proust to 'Gilligan's Island,' 'Mystery Science Theater 3000' fuses superb, clever writing with wonderfully terrible B-grade movies".[95] In 1994 and 1995, the show was nominated for the Primetime Emmy Award for Outstanding Individual Achievement in Writing for a Variety or Music Program, but lost both times to *Dennis Miller Live*.[96] Every year from 1992 to 1997, it was also nominated for CableACE Awards.[97][98] Its DVD releases have been nominated for Saturn Awards in 2004, 2006 and 2007.

18.8.3 Influence

Through *MST3K*, many obscure films have been more visible to the public, and several have since been considered some of the worst films ever made and are voted into the Bottom 100 on the Internet Movie Database.[99] Of note is *Manos: The Hands of Fate*, which was riffed on by *MST3K* in its fourth season. *Manos* was a very low-budget film produced by Hal Warren, a fertilizer salesman at the time, taking on a dare from a screenwriter friend to show that anyone could make a horror film. The film suffered from numerous production issues due to its limited filming equipment, and many critics describe the result using a riff from *MST3K*, in that "every frame of this movie looks like someone's last-known photograph".[100] The *MST3K* episode featuring *Manos* was considered one of its most popular and best episodes, and brought *Manos* into the public light as one of the worst films ever produced. The film gained a cult following, and presently there is an effort to restore the film to high-definition quality from its original film reels.[101] *MST3K* also riffed on three films directed by Coleman Francis, *Red Zone Cuba*, *The Skydivers*, and *The Beast of Yucca Flats*, which brought awareness of Francis' poor direction and low-budget films, similar to that of Ed Wood.[102] *MST3K* also brought to the limelight lackluster works by Bert I. Gordon, primarily giant monster B-movies, that gained attention through the show, and many Japanese kaiju movies imported and dubbed through producer Sandy Frank, particularly those in the *Gamera* series.[20]

MST3K's riffing style to poke fun at bad movies, films, and TV shows, have been used in other works.[103] In 2003, the television series *Deadly Cinema*, starring Jami Deadly, debuted, which featured the cast making fun of bad movies, *MST3K*-style. In 2004, the ESPN Classic series *Cheap Seats*, debuted, which featured two brothers making fun of clips of old sporting events, *MST3K*-style, and is noteworthy for containing an episode in which *MST3K* cast members briefly appeared in a cameo to make fun of the hosts' own skits. In 2008, the internet and direct-to-DVD comedy series *Incognito Cinema Warriors XP*, debuted, which used the same "host segment-movie segment" format the show established, while featuring completely original characters and plot. *ICWXP* gained a similar cult following, even earning the praises of former *MST3K* host Michael J. Nelson.[104] In 2010, the television series *This Movie Sucks!* (and its predecessor *Ed's Nite In*), starring Ed the Sock and

co-hosts Liana K and Ron Sparks, debuted. It features the cast making fun of bad movies. Creator Steven Kerzner, however, was quick to point out that *MST3K* was not "the creator of this kind of format, they're just the most recent and most well-known".[105] In 2011, the theater silhouette motif was parodied by golf commentator and talk show host David Feherty in an episode of *Feherty*. He is shown sitting in front of a large screen and "riffing" while viewing footage of golfer Johnny Miller and is joined in the theater by his stuffed rooster (Frank) and his gnome statue (Costas).

Further, the riffing style from *MST3K* is considered part of the influence for DVD commentaries and successful YouTube reviewers and Let's Play-style commentators.[10] DVD releases for both *Ghostbusters* and *Men in Black* used a similar format to Shadowrama for an "in-vision" commentary features.[106][107] The concept of social television, where social media is integrated into the television viewing experience, was significantly influenced by *MST3K*.[108] This social media practice of live-tweeting riffs and jokes on broadcast shows, such as for films like *Sharknado*, has its roots in *MST3K*.[11][15][109][110] The *MST3K* approach has inspired Internet movie critics to create comedic movie reviews approaches, such as through RedLetterMedia and Screen Junkies which are more considered more than just snarking on the movie but aim to help the viewer understand film and story techniques and their flawed use in poorly-received films.[111]

Public performances of live riffing have been hosted by various groups in different cities across the U.S. and Canada, including Cineprov (Atlanta, Georgia), Master Pancake Theater (Austin, TX), Counterclockwise Comedy (Kansas City, Missouri), FilmRoasters (Richmond, Virginia), Moxie Skinny Theatre 3000 (Springfield, Missouri), Riff Raff Theatre (Iowa City, Iowa), Twisted Flicks (Seattle, Washington), and Turkey Shoot (Metro Cinema at the Garneau, Edmonton, Alberta, Canada).[112][113][114] Canadian sketch comedy group Loading Ready Run produces the show *Unskippable* for The Escapist website, which applies the MST3K premise to video game cut scenes.

18.8.4 Fandom

MST3K, broadcasting during the emergence of the Internet for public use, developed a large fan base during its broadcast which continues to thrive since.[6] The show had already had its postal-based fan club, which people could write into and which some letters and drawings read on subsequent episodes, and the producers encouraged fans to share recordings with others.[6] At its peak, the "MST3K Fan Club" had over 50,000 members,[29] and Best Brains were receiving over 500 letters each week.[4] Fans of the show generally refer to themselves as "MSTies".[6] Usenet newsgroups rec.arts.tv.mst3k.misc and rec.arts.tv.mst3k.announce were established in the mid-1990s for announcements and discussions related to the show.[115][116][117] A type of fan fiction called MiSTings, in which fans would add humorous comments to other, typically bad, fan fiction works, was popular on these groups.[118] The fan-run website *Satellite News* continues to track news and information about the show and related projects from its cast members.[119] Another fan site, *The Annotated MST*, attempts to catalog and describe all the obscure popular culture references used in a given episode.[67]

In addition to the show's fandom, a number of celebrities have expressed their love for the show. One of the earliest known celebrity fans was Frank Zappa, who went so far as to telephone Best Brains, calling *MST3k* as "the funniest fucking thing on TV" according to Beaulieu.[6] Zappa became a friend of the show, and following his death episode 523 was dedicated to him. Other known celebrities fans include Al Gore, Penn Jillette, and Patton Oswalt.[6]

There were two official fan conventions in Minneapolis, run by the series' production company (called "ConventioCon ExpoFest-A-Rama" (1994) and "ConventioCon ExpoFest-A-Rama 2: Electric Bugaloo" (1996). At least 2,500 people attended the first convention.[6]

18.9   Related post-show projects

Mystery Science Theater 3000s *Mike Nelson (left) and Kevin Murphy, at "Exoticon 1" convention panel in Metairie, Louisiana, November 1998*

Main articles: Rifftrax and Cinematic Titanic

The various cast and crew from the show's broadcast run have continued to produce comedy works following the show. Two separate projects were launched that specifically borrowed on the theme of riffing on bad movies. After the short-lived *The Film Crew* in 2006, Nelson started *Rifftrax*, providing downloadable audio files containing *MST3K*-style

riffs that the viewer would synchronize to their personal copy of a given popular movie (such as *Star Wars: Episode I*); this was done to avoid copyright and licensing issues with such films. *Rifftrax*'s cast expanded to include Murphy and Corbett along with occasional guest stars, and were able to use a wider range of films, including films and shorts in the public domain, and films which they could get the license to stream and distribute. In addition, they launched production of *Rifftrax Live* shows for various films, where they performed their riffing in front of a live audience that was simultaneously broadcast to other movie theaters across the country and later made available as on-demand video. As of 2016, *Rifftrax* continues to offer new material and shows. As part of a tribute to their roots, *Rifftrax* has performed some works that previously appeared on *MST3K*, including *Manos: the Hands of Fate*, *Santa Claus*, and *Time Chasers*.

Similarly, Hodgson, after some experimental creative works such as *The TV Wheel*,[6] started *Cinematic Titanic* with Beaulieu, Weinstein, Conniff, and Pehl in 2007. Like *MST3K*, the five riffed on bad movies they were able to acquire the licenses for (including *Santa Claus Conquers the Martians*), which then were distributed through on-demand video and streaming options. They later did a number of live shows across the United States, some which were made available for digital demand. Production of *Cinematic Titanic* was shut down in January 2014.[120]

Other related projects by the *MST3K* crew following the show's end include: In 2000, most of the cast of the Sci-Fi era of the show collaborated on a humor website, Timmy Big Hands, that closed in 2001.

In 2001, Mike Nelson, Patrick Brantseg, Bill Corbett, Kevin Murphy and Paul Chaplin created *The Adventures of Edward the Less*, an animated parody of J. R. R. Tolkien's *The Lord of the Rings* and others in the fantasy genre, with additional vocals by Mary Jo Pehl and Mike Dodge, for the Sci Fi Channel website.[121]

In 2008, Bill Corbett and fellow writer Rob Greenberg wrote the screenplay for *Meet Dave*, a family comedy starring Eddie Murphy about a tiny *Star Trek*-like crew operating a spaceship that looks like a man. The captain of the crew and the spaceship were both played by Murphy. Originally conceived as a series called *Starship Dave* for SciFi.com, it was dropped in favor of *Edward the Less*. The script (along with the title) were changed drastically by studio executives and other writers, although Corbett and Greenberg received sole screenwriter credit.[6][122]

In 2010, Trace Beaulieu, Frank Conniff, Joel Hodgson, Mary Jo Pehl, Josh Weinstein, Beth McKeever and Clive Robertson voiced characters for *Darkstar: The Interactive Movie*, a computer game created by J. Allen Williams.

In 2013, Frank Conniff and animation historian Jerry Beck

debuted *Cartoon Dump*,[123] a series of classically bad cartoons, which are also occasionally performed live.[124]

In 2015, Trace Beaulieu and Frank Conniff began performing together as "The Mads", riffing movies at live screenings across the U.S.[125]

18.9.1 Reunions

In 2008, to commemorate the show's 20th anniversary, the principal cast and writers from all eras of the show reunited for a panel discussion at the San Diego Comic-Con International, which was hosted by actor-comedian Patton Oswalt. The event was recorded and included as a bonus feature on the 20th Anniversary DVD release via Shout! Factory. Also that year, several original *MST3K* members (including Joel Hodgson, Trace Beaulieu and Frank Conniff) reunited to shoot a brief sketch to be included on the web-exclusive DVD release of *The Giant Gila Monster*.[126] The new disc was added to Volume 10 of the "MST3K Collection" DVD boxed set series, replacing the *Godzilla vs. Megalon* disc which could no longer be sold due to copyright conflicts. The new package was sold under the name "Volume 10.2", and the sketch was presented as a seminar to instruct consumers on how to "upgrade" their DVD set, which merely consists of "disposing" of the old disc and inserting the new one.

In 2013, Joel Hodgson and Trace Beaulieu reprised their roles as Joel Robinson and Crow T. Robot for cameo appearances in the fourth season of *Arrested Development*.[127]

As part of its live show events for 2016, Rifftrax will host an open *MST3K* reunion at a live show in Minneapolis in June 2016, to include Hodgson, Bridget Nelson, Pehl, Conniff, and Beaulieu, and potentially other former cast and crew, as well as Jonah Ray from the revival. The gathered cast will riff on educational shorts as part of the event.[128][129]

18.10 See also

- List of films considered the worst

- Horror host

18.11 References

[1] "MST3k: Technical Specifications". *IMDb*. Retrieved 3 April 2015.

[2] "Company Credits". *IMDb*. Retrieved 5 April 2015.

[3] Hirsh, Mark (November 27, 2013). "About 'Mystery Science Theater,' A Bold Declaration. It's Bold!". *NPR*. Retrieved December 3, 2015.

[4] Winslow, Harriet (October 17, 1993). "'D' FLICKS, TWO 'BOTS, NEW HOST". *Washington Post*. Retrieved May 20, 2016.

[5] Itzkoff, Dave (November 9, 2008). "The Show That Turned the Mockery Into the Message". *New York Times*. Retrieved December 2, 2015.

[6] Referty, Brian (April 22, 2014). "Mystery Science Theater 3000: The Definitive Oral History of a TV Masterpiece". *Wired*. Retrieved December 2, 2015.

[7] Kangas, Chaz (November 27, 2013). "Talking Mystery Science Theater 3000's 25th Anniversary with Creator Joel Hodgson". *Village Voice*. Retrieved December 2, 2015.

[8] Dube, Jonathan; Perkins, Will (December 19, 2011). "Mystery Science Theater 3000 (1998)". *Art of the Title*. Retrieved December 2, 2015.

[9] Davis, Lauren (November 4, 2012). "How did MST3K pick those terrible, terrible movies?". *io9*. Retrieved December 2, 2015.

[10] Corlis, Richard (August 28, 2010). "Mystery Science Theater 2010: Riffer Madness!". *Time*. Retrieved December 2, 2015.

[11] Truitt, Brian (November 23, 2013). "Sunday Geekersation: 25 years of Joel Hodgson's 'MST3K'". *USA Today*. Retrieved December 3, 2015.

[12] Sloan, Will (August 16, 2012). "'You Can't Just Be The Voice Of Generic Sarcasm': The Art Of Movie Riffing". *NPR*. Retrieved December 2, 2015.

[13] Abrams, Simon (December 10, 2015). "Joel Hodgson: MST3K Isn't About Ridicule-It's a "Variety Show Built on the Back of a Movie"". *LA Weekly*. Retrieved December 14, 2015.

[14] Trace Beaulieu; et al. (1996). *The Mystery Science Theater 3000 Amazing Colossal Episode Guide* (1st ed.). New York: Bantam Books. pp. 153–159. ISBN 9780553377835.

[15] Wood, Jennifer (December 22, 2015). "'MST3K' Returns: Joel Hodgson on Resurrecting the Cult TV Show". *Rolling Stone*. Retrieved December 23, 2015.

[16] Trace Beaulieu; et al. (1996). *The Mystery Science Theater 3000 Amazing Colossal Episode Guide* (1st ed.). New York: Bantam Books. p. 145. ISBN 9780553377835.

[17] Hodgson, Joel (May 7, 2016). "The Return of... Cambot!". *Kickstarter*. Retrieved May 7, 2016.

[18] Philps, Keith (November 3, 2008). "The Mystery Science Theater 3000 reunion interview: Joel Hodgson, Trace Beaulieu, and Jim Mallon". *A.V. Club*. Retrieved December 2, 2015.

[19] "A Guy Named AJ : Launching Cinematic Titanic". *Star-Wars.com*. Archived from the original on 2007-12-02. Retrieved 2007-11-12.

[20] Vorel, Jim (August 10, 2015). "The 25-Episode History of Mystery Science Theater 3000". *Paste*. Retrieved December 1, 2015.

[21] Hilty, Wyn (August 5, 1999). "MST3K's Legacy Will Live On". *OC Weekly*. Retrieved December 3, 2015.

[22] Shales, Tom (November 27, 1991). "'MST3K' MEANS FINE TELEVISION". *Washington Post*. Retrieved May 20, 2016.

[23] Henry, Brian. "MST3K FAQ -- West Brains: Aliens in L.A.". MST3K Info Club. Archived from the original on 2007-04-14. Retrieved 2007-05-24.

[24] Phipps, Keith (1999-04-21). "Joel Hodgson". The Onion A.V. Club. Archived from the original on October 10, 2007. Retrieved 2007-07-12.

[25] Shales, Tom (December 29, 2008). "Web TV Review: Classic 'Mystery Science Theater 3000' Cast Returns for 'Cinematic Titanic'". *Washington Post*. Retrieved December 1, 2015.

[26] A.V. Club Staff (November 4, 2013). ""After a couple of years no one will even remember": 16 pop-culture windows into the world of 1993". *A.V. Club*. Retrieved December 14, 2015.

[27] Svetkey, Benjamin (December 15, 1995). "R.I.P. 'Mystery Science Theater 3000'". *Entertainment Weekly*. Retrieved January 15, 2016.

[28] O'Connell, Michael (January 13, 2016). "Viacom's Doug Herzog on MTV's Future, Keeping Amy Schumer on Comedy Central". *Hollywood Reporter*. Retrieved January 15, 2015.

[29] Karlen, Neal (February 2, 1997). "The Thing That Mocked The Movies". *New York Times*. Retrieved May 20, 2016.

[30] Patrick Brantseg Celebrity | TVGuide.com

[31] Muffin, Lawrie (June 26, 1996). "TV Notes - More Goofiness for Misties". *New York Times*. Retrieved May 20, 2016.

[32] Keepnews, Peter (July 25, 1999). "TELEVISION/RADIO; Think What They'd Do With 'Titanic'". *New York Times*. Retrieved May 20, 2016.

[33] Newman, Andrew Adam (May 6, 2007). "'MST3K': The Final Frontier". *New York Times*. Retrieved December 3, 2015.

[34] VanDerWuff, Todd (November 13, 2015). "Exclusive: Mystery Science Theater 3000 creator Joel Hodgson explains why it's time for a comeback". *Vox*. Retrieved November 13, 2015.

[35] Khatchatourian, Maane (April 22, 2014). "'Mystery Science Theater 3000' Reboot May Come Online in Not-Too-Distant Future". *Variety*. Retrieved December 2, 2015.

[36] Wagmeister, Elizabeth (November 10, 2015). "'Mystery Science Theater 3000' Acquired by Shout! Factory, Kickstarter Launched for New Season". *Variety*. Retrieved November 13, 2015.

[37] Anderson, Kyle. "Mystery Science Theater 3000 is returning". *Entertainment Weekly*. Retrieved 10 November 2015.

[38] Sagers, Aaron (January 29, 2016). "Back to the Satellite of Love: Joel Hodgson talks the behind the scenes journey back to Mystery Science Theater 3000". *Blastr*. Retrieved February 23, 2016.

[39] Vorel, Jim (November 16, 2015). "Joel Hodgson on the MST3k Revival, Criticism and #BringBackMST3k". *Paste*. Retrieved December 10, 2015.

[40] Grinberg, Emanuella (December 12, 2015). "'Mystery Science Theater 3000' revival sets new Kickstarter record". *CNN*. Retrieved December 14, 2015.

[41] Frank, Allegra (November 17, 2015). "Mystery Science Theater 3000 revival introduces a new host". *Polygon*. Retrieved November 17, 2015.

[42] Frank, Allegra (December 9, 2015). "Mystery Science Theater 3000 revival ending campaign with star-studded telethon". *Polygon*. Retrieved December 14, 2015.

[43] Osborn, Alex (December 12, 2015). "MST3K Breaks Veronica Mars' Record for Crowdfunding Film/Video". *IGN*. Retrieved December 12, 2015.

[44] Hughes, William (December 12, 2015). "MST3K breaks Kickstarter records, secures 14 new episodes". *A.V. Club*. Retrieved December 12, 2015.

[45] Vorel, Jim (December 11, 2015). "MST3k Becomes the Biggest Crowd-Funded Video Project in Internet History". *Paste*. Retrieved December 11, 2015.

[46] Vorel, Jim (November 23, 2015). "Felicia Day Officially Joins the Cast of the MST3k Reboot". *Paste*. Retrieved November 23, 2015.

[47] Rife, Katie (November 30, 2015). "Patton Oswalt is TV's Son of TV's Frank on the new Mystery Science Theater 3000". *The A.V. Club*. Retrieved November 30, 2015.

[48] Otterson, Joe (May 20, 2016). "'Mystery Science Theater 3000' Reboot Adds Former 'Daily Show' Head Writer". *The Wrap*. Retrieved May 20, 2016.

[49] Watts, Steve (December 9, 2015). "Dan Harmon to Script Mystery Science Theater 3000 Reboot". *IGN*. Retrieved December 9, 2015.

[50] Cecchini, Mike (December 12, 2015). "MST3K Reboot News: 14 Episodes Funded, Full Details Here". *Den of Geek*. Retrieved December 12, 2015.

[51] Hughes, William (December 9, 2015). "Dan Harmon and Justin Roiland are both writing for the new MST3K". *A.V. Club*. Retrieved December 10, 2015.

[52] Van Allen, Eric (December 10, 2015). "Jack Black, Jerry Seinfeld, Mark Hamill and More Could Appear in MST3K Revival". *Paste*. Retrieved December 10, 2015.

[53] Something's Not Right About the New, Crowdfunded 'Mystery Science Theater 3000' – Flavorwire

[54] Brown, Tracy (November 10, 2015). "'Mystery Science Theater 3000' launches Kickstarter campaign to reboot the series". *Los Angeles Times*. Retrieved November 13, 2015.

[55] https://www.facebook.com/mikeatrifftrax/posts/ 1167399969942922

[56] Burlingame, Russ (January 28, 2016). "Rifftrax's Kevin Murphy and Tommy Wiseau Talk The Room". *Comicbook.com*. Retrieved January 28, 2016.

[57] Frank Conniff on Twitter: ".@ShoutFactory is cutting us in, financially at least. They are a cool company. https://t.co/ gHvgOp6Gh7"

[58] Alter, Ethan (February 10, 2016). "'MST3K' Update: Joel Hodgson Shares New Details About the Reboot". *Yahoo! TV*. Retrieved February 20, 2016.

[59] Adams, Erik (November 27, 2013). "Twenty-five years on, there's reason to keep MST3K circulating". *A.V. Club*. Retrieved December 1, 2015.

[60] "Season 'Zero': KTMA-TV Channel 23 1988-1989". *Mystery Science Theater 3000: The Unofficial Episode Guide*. Satellite News. Retrieved 2007-01-26.

[61] "Jim Mallon interview". *Satellite News Interview of Jim Mallon*. Satellite News. Retrieved 2008-07-19.

[62] "Turkey Day, an 'MST3K' tradition". CNN. November 24, 2011. Retrieved February 23, 2016.

[63] Anderson, Kyle (2013-11-18). "MSTies rejoice! The 'Mystery Science Theater 3000' tradition Turkey Day is back! -- EXCLUSIVE". *Entertainment Weekly*. Retrieved 2013-11-18.

[64] Anderson, Kyle (2014-11-03). "EXCLUSIVE: JOEL HODGSON CURATING ANOTHER MST3K TURKEY DAY STREAMING MARATHON". The Nerdist. Retrieved 2014-11-03.

[65] McCormick, Luke (November 24, 2015). "How Mystery Science Theater 3000 Came Back to Life". *Vulture*. Retrieved December 1, 2015.

[66] Duckworth, Courtney (September 19, 2014). "Where Do I Start With Mystery Science Theater 3000?". *Slate*. Retrieved December 2, 2015.

[67] Hughes, William (October 22, 2015). "Watch an annotated MST3K episode for free on YouTube". *A.V. Club*. Retrieved December 3, 2015.

[68] "Shout! Factory's free video service is a cult movie and TV fan's dream". *Engadget*. February 5, 2015. Retrieved December 2, 2015.

[69] Cecchini, Mike (November 9, 2015). "Mystery Science Theater 3000 and RiffTrax Team Up for MST3K Mondays". *Den of Geek*. Retrieved December 2, 2015.

[70] Beaulieu, Trace; et al. ""The Mystery Science Theater Hour"". *The Mystery Science Theater 3000 Amazing Colossal Episode Guide*. p. 111.

[71] Adams, Eric (April 4, 2014). "Mystery Science Theater 3000 reruns airing on Retro TV this summer". *A.V. Club*. Retrieved December 2, 2015.

[72] MST3K on PBS Update: Currently airing in Lubbock, TX, and coming soon to Atlanta. : MST3K

[73] Mystery Science Theater 3000 | WTTW Chicago

[74] Lambert, David, ed. (January 22, 2008). "New DVD Release for MST3K: The Movie...at last!". "DVD News" (column), *TV Guide*. Archived from the original on June 3, 2009. Retrieved 2008-02-03.

[75] Beaulieu, Trace; et al. *The Mystery Science Theater 3000 Amazing Colossal Episode Guide*. p. 64.

[76] Beaulieu, Trace; et al. "Forward About Kurt Vonnegut". *The Mystery Science Theater 3000 Amazing Colossal Episode Guide*. pp. xi–xiii.

[77] "PlayStation Perfect Guide". Game Rave. Archived from the original on July 8, 2007. Retrieved 2007-06-02.

[78] *MSNBC* article: "Ex 'MST3K' stars, writers fill hole left by show". webcitation archive link

[79] "25 Top Cult Shows Ever!". *TV Guide* (May 30 – June 5, 2004): 32. ISSN 0039-8543.

[80] "TV Guide Names the Top Cult Shows Ever". June 29, 2007. Retrieved August 24, 2007.

[81] Poniewozik, James (6 September 2007). "The 100 Best TV Shows of All-*Time*". *Time*. Retrieved 4 March 2010.

[82] "25 Best Cult TV Shows from the Past 25 Years". *Entertainment Weekly*. August 3, 2012. p. 37.

[83] "Part 14: Battles on Many Fronts (1996)". *The Almost but Still Not Quite Complete History of MST3K*. Satellite News. Retrieved 2006-08-17.

[84] Cornell, Chris (2010-11-25). "Turkey Day Surprise from Shout". Retrieved 2011-01-17.

[85] Ward, Brian (2010-11-29). "Shout Factory Community". Archived from the original on January 1, 2011. Retrieved 2011-01-17.

[86] Finley, Stephen F. (June 25, 1999). "512 - Mitchell". *Daddy-O's Drive-In Dirt*. Satellite News. Retrieved 2006-08-17.

[87] Chandler, Rick. "MST3K Touches Down For Good". *Impression Magazine*. Reprinted by MSTies Anonymous. Archived from the original on 2006-06-18. Retrieved 2006-08-17.

[88] Cavanaugh, Maureen (2006-08-30). "Host of Mystery Science Theater 3000 moves to San Diego" (MP3). *These Days*. KPBS-FM. Retrieved 2006-09-13.

[89] Sloane, Rick (2006). Interview with Jonah Falcon. The Jonah Falcon Show. MNN. New York City. Missing or empty |title= (help);

[90] Borntreger, Andrew (February 2, 2008). "Interview with Rick Sloane". *BadMovies.org*. Retrieved 2008-03-11.

[91] Jeff Lieberman, director. (1976). "Director's Commentary", *Squirm* (NTSC) [DVD], MGM. Released August 26, 2003.

[92] "An Interview with Fiveson & Schriebman". The Mystery Science Theater 3000 Review. Retrieved 2006-08-17. Original discussion was started under the thread "Interview with Robert Fiveson" on Proboards on July 29, 2005.

[93] Beaulieu, Trace; et al. "Miles O'Keefe: A Tribute". *The Mystery Science Theater 3000 Amazing Colossal Episode Guide*. p. 37.

[94] "An Interview With David Giancola". The Mystery Science Theater 3000 Review. *c*. May 22, 2005. Archived from the original on September 4, 2007. Retrieved 2006-08-17. Check date values in: |date= (help) Date is based on information on the discussion thread "David Giancola Interview".

[95] "The Peabody Awards". Grady College of Journalism and Mass Communication at the University of Georgia. Archived from the original on 2007-07-08. Retrieved 2007-07-10.

[96] "Mystery Science Theater 3000". Academy of Television Arts & Sciences. Retrieved 3 August 2011.

[97] Weiner, Robert G.; Barba, Shelley E., eds. (2011). *In the Peanut Gallery with Mystery Science Theater 3000 Essays on Film, Fandom, Technology and the Culture of Riffing*. Jefferson: McFarland & Co., Publishers. p. 7. ISBN 9780786485727. Retrieved 5 April 2015.

[98] Lavery, David (2015). *The Essential Cult TV Reader - Essential Readers in Contemporary Media and Culture*. University Press of Kentucky. pp. 181–182. ISBN 9780813150208. Retrieved 5 April 2015.

[99] Dillon-Trenchard, Pete (July 4, 2012). "Looking back at Mystery Science Theatre 3000". *Den of Geek*. Retrieved December 3, 2015.

[100] Blevins, Joe (September 25, 2015). "Read this: The battle over the infamous cult classic Manos: The Hands of Fate". *A.V. Club*. Retrieved December 3, 2015.

[101] Rossen, Jake (September 24, 2015). "The Battle Over the Worst Movie Ever". *Playboy*. Retrieved December 3, 2015.

[102] Vorel, Jim (December 2015). "Coleman Francis: The Real Worst Director in Film History". *Paste*. Retrieved December 3, 2015.

[103] Burube, Chris (February 1, 2013). "That Awful Film, Incredibly, Has a Following". *New York Times*. Retrieved May 20, 2016.

[104] Rikk Wolf on How Incognito Cinema Warriors makes Terrible Movies Better Archived March 18, 2011, at the Wayback Machine.

[105] Mohawk students help bring Ed back to TV Hamilton MountainNews, May 27, 2010 (Article by Gord Bowes)

[106] A.V. Club Staff (November 10, 2010). "Now with extra farts! 25 1/2 gimmicky DVD commentary tracks". *A.V. Club*. Retrieved December 3, 2015.

[107] Gaudiosi, John (June 22, 2008). "Men in Black (Blu-ray Review)". *Home Media Magazine*. Retrieved December 2, 2015.

[108] Cesar, Pablo (2009). *Social Interactive Television: Immersive Shared Experiences and Perspectives: Immersive Shared Experiences and Perspectives*. IGI Global. p. 3. ISBN 1605666572.

[109] Garber, Megan (November 11, 2015). "Jumping the Snark: The Timely Return of Mystery Science Theater 3000". *The Atlantic*. Retrieved December 8, 2015.

[110] Alston, Joshua (July 20, 2015). "Syfy's Sharknado series proves schlock cinema belongs on TV, not in theaters". *A.V. Club*. Retrieved December 13, 2015.

[111] Worthington, Clint (April 19, 2016). "The Enduring Legacy of Mystery Science Theater 3000". *Consequence of Sound*. Retrieved April 19, 2016.

[112] Raspberry Brothers and the Many Spawn of MST3K

[113] FilmRoasters Fry at the Byrd

[114] The Sensational Saga of Mr. Sinus

[115] Godes, David; Dina Mayzlin (August 2003). "Using Online Conversations to Study Word of Mouth Communication" (PDF). pp. 10–11. Archived from the original (PDF) on April 8, 2011. Retrieved 15 September 2010. We found 169 different groups that contained messages about the shows in our sample ... Table 3 ... 20 Top Newsgroups in the Sample ... rec.arts.tv 9,649 ... rec.arts.tv.mst3k.mis 578

[116] Lieck, Ken (July 14, 1995). "The Information Dirt Road Marketing Your Band on the Net". *The Austin Chronicle*. Retrieved 15 September 2010. groups where obsessos of all types get together and exchange information about their favorite TV shows (news:alt.tv.brady-bunch, news:rec.arts. tv.mst3k)

[117] Werts, Diane (May 14, 1996). "A MSTie Farewell to Mike, The 'Bots and Bad Flicks". *Newsday* (Long Island, N.Y.). pp. B.53. new MST feature flick which just ended its NYC run and should hit the burbs soon check the Internet newsgroup rec.arts.tv.mst3k

[118] Murray, Noel (January 29, 2012). "Cheap Seats, "Superstars 1978"". *A.V. Club*. Retrieved December 2, 2015.

[119] Reed, Ryan (April 23, 2015). "'MST3K' Creator Hints at Online Reboot". *Rolling Stone*. Retrieved December 2, 2015.

[120] http://austin.com/articles/1013/ the-austincom-interview-mst3ks-mary-jo-pehl.html

[121] Murphy, Kevin. "Edward the Less Video Interview, Part 1". SciFi.com, 2001. Retrieved on 2009-02-02. Archived March 4, 2009, at the Wayback Machine.

[122] Corbett, Bill. "Meat, Dave?", *Rifftrax.com*, July 10, 2008. Retrieved on 2009-02-02.

[123] Cartoondump.com

[124] Ticket Sales - Cartoon Dump! at The Steve Allen Theater at The Center For Inquiry on Monday, January 23, 2012

[125] The Mads Are Back » The Mads are back! Mystery Science Theater 3000 stars Frank Conniff and Trace Beaulieu are taking movie riffing back onstage in their new live show

[126] "Joel & The 'Bots Return for Brief DVD Reunion". *Wired: Underwire blog*. 2008-01-30. Retrieved 2008-10-05.

[127] Bosch, Torie; Wade, Chris (June 3, 2013). "Arrested Development, Season 4". *Slate*. Retrieved March 23, 2016.

[128] Hughes, William (March 9, 2016). "Get Involved, Internet: RiffTrax is putting together an MST3K reunion show". *A.V. Club*. Retrieved March 23, 2016.

[129] Rife, Katie (April 1, 2016). "Rifftrax Exclusive: Joel Hodgson and Jonah Ray join the live MST3K reunion". *A.V. Club*. Retrieved April 1, 2016.

18.12   External links

- *Mystery Science Theater 3000* at the Internet Movie Database

- *Mystery Science Theater 3000* at TV.com

Chapter 19

Mystery Science Theater 3000 (Flash series)

Mystery Science Theater 3000, also referred to as **"The 'Bots Are Back!"** was an Internet cartoon created by Best Brains, Inc. It was inspired by BBI's original *Mystery Science Theater 3000* TV series, and was directed by former Executive Producer Jim Mallon. The series featured the robot characters from the original series in a variety of brief sketches taking place at an undetermined point during the original show's fictional timeline. However, no human character (Joel or Mike) was present.

19.1 Cast

- Paul Chaplin as Crow T. Robot
- James Moore as Tom Servo
- Jim Mallon as Gypsy

19.2 History

The animated version of MST3k was first announced on the Satellite News website (formerly BBI's official website) on October 29, 2007. The new cartoon was described as a "weekly series of animated adventures," and would debut as part of an all-new MST3k website. This new site would also feature content from the original series as well as a new online store. The website went live on November 5, 2007 along with the first installment of the animated series.

New episodes were scheduled to be posted every Monday. However, despite regular updates to other parts of the site, no new episodes of the cartoon have been posted since November 26, 2007. In June 2008, the website was redesigned and the cartoons removed for unspecified reasons; however the Flash likenesses of the characters were still present on some pages. On July 18, MST3k fansite *Satellite News* posted an interview with Jim Mallon, who explained that the cartoons cost more to produce than was initially estimated. He also stated that the existing cartoons

will return to the site, and expressed hope that the series may continue at some point in the future.[2] Animation studio Shad Petosky responded that the costs were low and flat rates that never changed from the original estimates, he suspects that the revenue was the problem and the mouse pads, post-it notes, and T-shirts being sold to pay for the show did not sell as fans did not like the poorly designed and written animated characters.[3]

19.3 Reactions

Initially, response to the new website on both of the major fan discussion boards was largely negative.[4] Michael J. Nelson, star and head writer of the original series, called the animated series "cute" but felt it was an "after-the-fact" idea.[4] Viewers of the flash series had said that the animation was poor and the voice actors did not fit with the characters. Later, the series was abandoned.

19.4 Episodes

1. **"Reel Livin'"**: Crow goes fishing and discusses the benefits of his "incredibly stable" kayak before being capsized by Servo on a jet ski.

2. **"Feels Like"**: Gypsy and Servo discuss actual temperature versus the wind chill factor. Servo then begins translating everything Gypsy says into "feels like..." statements, much to her annoyance.

3. **"Thanksgiving Clown"**: Servo dresses up as a clown, thinking it to be a Thanksgiving tradition. Crow consults an old World Book encyclopedia to verify.

4. **"Solitaire"**: Crow plays solitaire while Servo advises. Eventually Crow becomes fed up with Servo's pestering and leaves. Servo sees a fantastic move, but can't

play as his arms don't work. Desperate, he unsuccessfully tries to move the cards using telekinesis, but in the process breaks himself.

19.5 References

[1] "Mst3k.com Site Info". Alexa Internet. Retrieved 2013-01-25.

[2] http://www.mst3kinfo.com/satnews/brains/mallon.html

[3] http://forrestcrow.proboards47.com/index.cgi?board=deep&action=display&thread=11964&page=30#1203662913

[4] *MSNBC* article: "Ex 'MST3K' stars, writers fill hole left by show".

Chapter 20

Mystery Science Theater 3000 home video releases

Mystery Science Theater 3000 is an American TV show which aired between 1988 and 1999. This page summarizes video tape and DVD releases of episodes of the show. Episodes were initially released by Rhino Entertainment, with the rights later being purchased by Shout! Factory. Releases usually consist of boxsets of 4 episodes, although early releases consisted of just single episodes.

As of June 2016, **151** of the series' 176 non-KMTA-TV episodes have been released on home video.

20.1 VHS releases

Rhino Home Video released several episodes from the Comedy Central era on VHS from April 1996 to January 2001. As of 2004, all of the tapes are out of print, but all episodes originally released on home video have been released on DVD either as a single or part of a volume pack (except for *309 - The Amazing Colossal Man* due to infringement of the copyright in the original movie).

Best Brains has produced VHS tapes for independent sale through their info club newsletters. As of 2007, all of the tapes are now out of print, but most of the tapes have been released as a bonus feature on DVD releases.

A listing of video tapes released is listed below.

20.1.1 Best Brains

20.1.2 Rhino

20.2 DVD releases

20.2.1 Rhino

Beginning in March of 2000, Rhino started to release episodes of *MST3K* on DVD.

One MST3K volume pack (*Volume 10*) was discontinued 2 months after its initial release due to copyright issues with *Godzilla Vs. Megalon*. The volume pack was reissued by Rhino *Volume 10.2*, with a new episode (*The Giant Gila Monster*).

As of 2014, all Rhino releases have been discontinued, due to the company no longer having the rights to distribute MST3K on home video.

A complete listing of releases is shown in the table below. Any DVDs listed as "Out of print" are currently no longer available on the Rhino (for the Rhino DVDs), Shout! Factory (for the Shout DVDs), or official Mystery Science Theater 3000 websites (for all DVDs).

20.2.2 Shout! Factory

In January 2008, Best Brains transferred the worldwide home entertainment and digital download license for *MST3K* from Rhino to Shout! Factory.

Item discontinued and no longer in print.

Re-release of an item previously released by RHINO Home Video.

20.2.3 Best Brains

Best Brains' DVDs originally from their fanclub, now on MST3K.com .

20.3 Mystery Science Theater 3000: The Movie

The feature film *Mystery Science Theater 3000: The Movie* was released on VHS and laserdisc in 1997 by MCA/Universal Home Video. Image Entertainment released it on DVD in 1998, but was taken out of print by MCA/Universal Home Video in 2000. *The Movie* was rereleased by Universal on May 6, 2008, and was released again in a Blu-ray/DVD combo pack by Shout! Factory on September 3, 2013.

20.4 References

[1] https://www.shoutfactory.com/film/sci-fi/mst3k-20th-anniversary-edition-collector-s-edition-tin

[2] https://www.shoutfactory.com/film/sci-fi/mst3k-20th-anniversary-edition-standard-edition

[3] https://www.shoutfactory.com/film/action-adventure/mst3k-volume-xiv

[4] https://www.shoutfactory.com/film/film-comedy/mst3k-volume-xv

[5] https://www.shoutfactory.com/film/film-comedy/mst3k-volume-xvi-collector-s-edition

[6] https://www.shoutfactory.com/film/film-comedy/mst3k-volume-xvi-standard-edition

[7] https://www.shoutfactory.com/film/action-adventure/mst3k-volume-xvii

[8] https://www.shoutfactory.com/film/sci-fi/mst3k-volume-xviii

[9] https://www.shoutfactory.com/film/sci-fi/mst3k-volume-xix-collector-s-edition

[10] https://www.shoutfactory.com/film/sci-fi/mst3k-volume-xix-standard-edition

[11] https://www.shoutfactory.com/film/sci-fi/mst3k-beginning-of-the-end

[12] https://www.shoutfactory.com/film/film-comedy/mst3k-the-incredibly-strange-creatures-who-stopped-living-became-mixed-up-zombies

[13] https://www.shoutfactory.com/film/martial-arts/mst3k-volume-xx

[14] https://www.shoutfactory.com/film/film-western/mst3k-gunslinger

[15] https://www.shoutfactory.com/film/film-drama/mst3k-hamlet

[16] https://www.shoutfactory.com/film/action-adventure/mst3k-volume-xxi-mst3k-vs-gamera-collector-s-edition-tin

[17] https://www.shoutfactory.com/film/film-comedy/mst3k-red-zone-cuba

[18] https://www.shoutfactory.com/film/film-comedy/mst3k-the-unearthly

[19] https://www.shoutfactory.com/film/film-horror/mst3k-manos-the-hands-of-fate-special-edition

[20] https://www.shoutfactory.com/film/film-comedy/mst3k-the-touch-of-satan

[21] https://www.shoutfactory.com/film/film-comedy/mst3k-the-atomic-brain

[22] https://www.shoutfactory.com/film/sci-fi/mst3k-volume-xxii

[23] https://www.shoutfactory.com/film/film-comedy/mst3k-the-wild-world-of-batwoman

[24] https://www.shoutfactory.com/film/film-drama/mst3k-girl-in-gold-boots

[25] https://www.shoutfactory.com/film/sci-fi/mst3k-volume-xxiii

[26] https://www.shoutfactory.com/film/sci-fi/mst3k-volume-xxiv

[27] https://www.shoutfactory.com/film/film-crime/mst3k-volume-xxv

[28] https://www.shoutfactory.com/film/sci-fi/mst3k-volume-xxvi

[29] https://www.shoutfactory.com/film/film-comedy/mst3k-volume-xxvii

[30] https://www.shoutfactory.com/film/film-horror/mst3k-25th-anniversary-edition-collector-s-edition-tin

[31] https://www.shoutfactory.com/film/action-adventure/mst3k-volume-xxix

[32] https://www.shoutfactory.com/film/film-horror/mst3k-volume-xxx

[33] https://www.shoutfactory.com/film/action-adventure/mst3k-volume-xxxi-the-turkey-day-collection-collector-s-edition-tin

[34] https://www.shoutfactory.com/film/film-comedy/mst3k-volume-xxxii

[35] https://www.shoutfactory.com/film/film-crime/mst3k-volume-xxxiii

[36] https://www.shoutfactory.com/film/film-comedy/mst3k-volume-i

[37] https://www.shoutfactory.com/film/action-adventure/mst3k-volume-xxxiv

[38] https://www.shoutfactory.com/film/action-adventure/
      mst3k-volume-xxxv

[39] https://www.shoutfactory.com/film/action-adventure/
      mst3k-volume-ii

[40] https://www.shoutfactory.com/film/sci-fi/
      mst3k-volume-xxxvi

Chapter 21

Mystery Science Theater 3000: The Movie

Mystery Science Theater 3000: The Movie is a 1996 American comedy film and a film adaptation of the television series *Mystery Science Theater 3000*, produced and set between seasons 6 and 7 of the show. It was distributed by Gramercy Pictures[1] and produced by Best Brains and Universal Studios.

The filmmakers dub a new comic narrative over the 1955 science fiction film *This Island Earth*, editing out approximately twenty minutes of the original film.

21.1 Plot

For the plot of the film-within-the-film, see This Island Earth

The film opens with mad scientist Dr. Clayton Forrester, working from an underground laboratory, explaining the premise of the film (and associated TV series). Mike Nelson and the robots Crow T. Robot and Tom Servo, along with Gypsy, are aboard the Satellite of Love high in Earth's orbit, when Forrester forces them to watch the film *This Island Earth* to break their wills; as per the television show, Mike, Crow, and Tom riff the film as it airs.

The film-riffing scenes are book-ended and interspersed with short sketches. Prior to the film, Crow attempts to dig through the ship's hull to return to Earth. After the filmstrip breaks and Forrester reloads it, Crow and Tom dare Mike to drive the Satellite himself, but ends up crashing into the Hubble Space Telescope; Mike then tries to repair the Hubble using the Satellite's manipulator arms, MANOS, but instead further damages the unit before Gypsy takes over. Some time into the film, Tom reveals that he has an interocitor like that used in *This Island Earth*. The gang tries to use Tom's device to return to Earth, but instead contact a Metalunan (the alien race from the film) who is unable to help them to figure out how to use it correctly but does accidentally repeatedly zap Tom's head with a laser beam. The contact is broken by Forrester, who also has an interoc-

itor, and he zaps the group to encourage them back to the theater.

After the film, Mike, Crow, and Tom are far from broken, celebrating in various Metaluna ways. Forrester, furious at his failure, attempts to use his own interocitor to harm Mike and the others, but only succeeds in transporting himself into the shower of the Metalunan previously seen. Mike and the robots briefly celebrate Forrester's disappearance before they realize they no longer have a way back to Earth without him. Upon this realization, the crew goes back to the theater to riff on the film's ending credits.

21.2 Cast

- Michael J. Nelson as Mike Nelson

- Trace Beaulieu as Crow T. Robot / Dr. Clayton Forrester

- Kevin Murphy as Tom Servo

- Jim Mallon as Gypsy

- John Brady as Benkitnorf

21.3 Production

The film was shot away from the Best Brains corporate headquarters and studio in Eden Prairie, Minnesota, at Energy Park Studios in St. Paul.[2]

Deleted scenes

- At the beginning of the film, it was originally planned to have a new version of the "MST3K Love Theme" by Dave Alvin.[3] However, the song was reduced to an instrumental version used for the end credits.

- To trim the film's duration, Gramercy ordered one of the host segments to be cut. In this scene, Mike and the bots hide out in the ship's storm shelter to avoid a meteor shower. The barrage of meteors threatens to damage the ship's oxygen supply, and Crow, Servo, and Gypsy rush to save Mike's life.

- The ending was also changed – originally, the film's final moments depicted Mike and the bots exacting revenge on Forrester by hooking up Servo's interocitor to the video feed from the Hexfield Viewscreen and sending a Metalunan mutant (played by MST3K prop man and toolmaster Jef Maynard) to strangle the mad scientist. At the end, Crow goes back to the basement to plan another escape attempt, this time armed with the chainsaw that he found in Servo's room earlier in the film.

- The new theme song, cut scene, and alternate ending were shown at the "Mystery Science Theater 3000 ConventioCon ExpoFest-O-Rama 2: Electric Bugaloo" in 1996, but were not included on home media releases until the Shout! Factory Collector's Edition.

21.4 Release

Mystery Science Theater 3000: The Movie was released on April 19, 1996. At its widest point of its North American theatrical release, the film screened in 26 cinemas.[1]

21.4.1 Box office

In its opening weekend, the film grossed $206,328. It went on to gross $1,007,306.[1]

21.4.2 Critical reception

The film received generally positive reviews from critics. On review aggregator website Rotten Tomatoes, the film holds an 80% rating, based on 54 reviews, with an average rating of 6.6/10. The site's consensus states: "*Mystery Science Theater 3000: The Movie* may be thin and uneven, but it's hilarious in enough of the right spots to do the show's big-screen transition justice."[4]

21.5 Home media

The film was released on VHS by MCA/Universal Home Video to rental outlets on October 1, 1996. The film was released for retail sales on April 8, 1997 on both VHS and Laserdisc formats. *MST3K: The Movie* was released on DVD in 1998 by Image Entertainment, as a discount title with an MSRP of $14.99.

Universal re-released the DVD on May 6, 2008. The film is in anamorphic widescreen, and includes an upgraded Dolby Digital 5.1 soundtrack, and English subtitles, a first for an MST3K DVD.

It was announced on June 7, 2013 that Shout! Factory would be releasing The Movie on a Blu-ray/DVD combo pack Collector's Edition. This release included, for the first time, the deleted scenes from the film.

TV airings

Besides being released on VHS, Laserdisc, and DVD, in recent years, the film has also been shown on the Starz and HBO television channels in the USA, and the film has been aired frequently in Europe on the Sky Movies channels.

21.6 See also

- List of films featuring space stations

21.7 References

[1] *Mystery Science Theater 3000: The Movie* at Box Office Mojo

[2] "*Mystery Science Theater 3000: The Movie* (1996) – Filming locations". *Internet Movie Database*. Amazon.com. Retrieved 2012-01-16.

[3] https://www.youtube.com/watch?v=3t0FvZ56fSM

[4] "Mystery Science Theater 3000: The Movie". *Rotten Tomatoes*. Flixster. Retrieved December 11, 2015.

21.8 External links

- *Mystery Science Theater 3000: The Movie* at the Internet Movie Database

- *Mystery Science Theater 3000: The Movie* at Box Office Mojo

- *Mystery Science Theater 3000: The Movie* at Rotten Tomatoes

Chapter 22

Observer (Mystery Science Theater 3000)

Observer (also known as **Brain Guy**) is a fictional character in the *Mystery Science Theater 3000* television series. He is played by Bill Corbett, and appears in the eighth through tenth seasons of the series.

Observer is a hyperintelligent, psychic alien from a planet of fellow aliens confusingly sharing the name "Observer" (the other two who appear in the series are played by Michael J. Nelson and Paul Chaplin). Supposedly, the Observers "evolved" beyond bodies into dark-green brains, contained in large Petri dishes (not unlike the Providers in the *Star Trek* episode "The Gamesters of Triskelion"). They are carried around by humanoid host bodies (controllable over a distance of up to 50 yards), rendering their abandonment of their original bodies rather pointless. (As the robot Gypsy points out, "Wouldn't it be more convenient to just keep your brains in your heads?") Thus Observer is, technically, only the brain which is being carried by the host body, but for all intents and purposes, he is considered a humanoid with brain separated from body. Observer joins the mad scientists ("The Mads") after his planet is inadvertently destroyed by Mike Nelson.

Like his colleagues Professor Bobo (Kevin Murphy) and Pearl Forrester (Mary Jo Pehl), Brain Guy is deeply dysfunctional. Unlike Bobo and Pearl, Brain Guy apparently has a considerable social and sexual life once the Mads returns to Earth in Seasons 9 and 10. He is also said to have, as Professor Bobo put it, "B.O." (body odor). Observer denies this, claiming that he doesn't have a body, although eventually he gives himself a sniff and admits that he does, in fact, stink.

Observer, after joining the Mads, usually ends up being the one who sends the movies to the Satellite of Love via his psychic abilities.

Observer, like his fellow Observers, claims to be omnipotent and omniscient, much like Q and his people from *Star Trek: The Next Generation*, but frequently fails to demonstrate these supposed abilities (he once stated that he was "not *that* omnipotent"). In one of his early appearances (Episode #806, *The Undead*), before his homeworld is accidentally destroyed by Mike Nelson, his fellow Observers test the rest of the cast to see if any of them deserve the right to become part of their kind. He is surprised when Tom Servo scores higher than him, leading to Observer being painfully punished.

Exposure to Pearl's autocratic manner appears to further degrade his powers over the course of the show, to the point that when Observer tries to punish Mike horribly, he only sends him a necktie. ("Don't you see what a *terrible* gift that is?") In the final episode, #1013, *Danger: Diabolik*, Pearl's playful dousing of his brain in Mountain Dew temporarily interferes with his speech and disables his gifts, allowing the Satellite of Love to crash to Earth.

22.1 External links

- Observer at the Internet Movie Database

Chapter 23

Pearl Forrester

Pearl Forrester is a character on the *Mystery Science Theater 3000* television series, played by Mary Jo Pehl. Forrester was the mother of Dr. Clayton Forrester (Trace Beaulieu). Initially devised as a guest character, Pearl would take on an increasingly important role in the series, first as a replacement supporting character for Dr Forrester and subsequently replacing him as the lead mad scientist.

Pearl's first appearance was in episode 607, *Bloodlust!*. Her character was featured in the opening and closing host segments when she paid a visit to Dr Forrester and his current assistant, TV's Frank. It became apparent that Frank had become a firm friend of Pearl's through a long correspondence, and Pearl was much more interested in spending time with him than her own son. At the end of the sixth season, TV's Frank died (and was guided to 'Second Banana Heaven' by 'Torgo The White', a reference to the reborn Gandalf the White from *Lord of the Rings*). Forrester moved in to help her son at the beginning of the seventh season, and became a regular cast member from that point until the end of the show's tenth and final season.

When Trace Beaulieu left at the end of the season, Pearl took over as the head Mad at the beginning of the eighth season. According to the backstory presented at that time, Pearl killed Dr. Forrester (after he had re-attained adulthood following his transformation into a *2001*-esque Space Baby), then vowed to "avenge his death" by continuing his experiments on Mike and the Bots; she had herself cryogenically frozen until the year 2525 (probably a reference to the 1960s song "In the Year 2525") and became the leader of the *Planet of the Apes*-like apes who now dominated Earth, at which point she (apparently) somehow drew the Satellite of Love's crew back to the ship so that she could send them bad movies. In the episode "The Deadly Mantis," the Earth was destroyed, when Professor Bobo and Dr. Peanut helped their mutant neighbors fix their thermonuclear device. From then until the end of the series, she was assisted by Professor Bobo, who often addressed her as "Lawgiver" (another *Planet of the Apes* reference), and Observer (aka Brain Guy). During the episode "Quest of the Delta Knights" she becomes frustrated with a significant lack of progress in the experiment and decides to do some role-reversal to see firsthand what is wrong with the experiment, switching places with Mike in the Satellite of Love while Mike takes her place in Castle Forrester. During the movie, in lieu of a smoke, she decides to suck on a mint and shares with the bots. She is later referred to as "Mintgiver" by the bots.

When the Satellite of Love returned to the present, Pearl and her lackeys followed and took up residence at Castle Forrester, ancestral home of the Forrester family, where she found records of a long line of Forresters who had performed experiments similar to those she and Dr. Forrester conducted on Mike and the Bots. Oddly, Pearl claims to be descended from these Forresters, indicating that Dr. Forrester used his mother's original surname, rather than that of his father (whoever that might be), which might suggest that he was born out of wedlock, or that the strong-willed Pearl refused to change her surname on marriage and insisted on her children bearing her name, or possibly that his father died before he was born. Indeed, Pearl has been married several times, with all of her husbands meeting gruesome fates on their honeymoons (it is implied that she was responsible for all their deaths):

- Chuck – became a prairie dog when he and Pearl visited a prairie dog colony in South Dakota; cause of death unrevealed

- Felipe – shot

- Maury – had hatpins shoved through his eyes right before the ceremony, presumably lingered long enough to in fact marry Pearl, then die on the honeymoon

- Wendell – shot

- Jerome – cause of death unrevealed (by her sinister tone of voice when she alludes to his death, it is implied he died in a particularly gruesome manner)

Which, if any, of these men was Dr. Forrester's father is unrevealed.

At the end of the series, Pearl accidentally used a new controller to send the Satellite of Love crashing back to Earth. She supposedly became the ruler of Qatar, where she vowed that her first item of business would be to insert a "U" in the nation's name. She is last seen with her lair devoid of furnishings (it is evident that they are nearly moved out), attempting to dance and sing "It's a Long Way to Tipperary" with her cronies by the light of a lone light bulb in a parody of the final episode of *The Mary Tyler Moore Show*. Her last words to Mike and the Bots before pulling the plug on Castle Forrester's observation equipment were, "Look, Nelson. Move on. I am."

23.1 External links

- Pearl Forrester at the Internet Movie Database

Chapter 24

Professor Bobo

Professor Bobo is a fictional character who appeared in the final three seasons of *Mystery Science Theater 3000*, a comedy television series that mocks B-movies. Played by Kevin Murphy (who also voiced and operated robot Tom Servo on the show), Bobo is a sapient, speaking gorilla from the year 2525, an homage to the film *Planet of the Apes*.

Professor Bobo is a curious mixture of eminent scientist, descended from a long line of supposedly respected apes, and of sensory-driven primate, often regressing to more primitive desires and actions. In the first three shows of Season 8, he leads an ape laboratory directed by the "Lawgiver" (*Planet of the Apes*) to continue the movie-watching "experiments" on Mike Nelson and his robots, again trapped in the Satellite of Love. In subsequent shows, Bobo travels with the Lawgiver, who turns out to be Pearl Forrester (inheriting the mad scientist role from her son, Dr. Clayton Forrester). Bobo ultimately becomes one of her henchmen after his planet is destroyed when Mike Nelson helps the apes and their new mutant friends activate an atomic bomb (a reference to *PotA* sequel *Beneath the Planet of the Apes*).

Despite his title of "Professor", Bobo demonstrated remarkably little intelligence after leaving Earth of 2525. He was revealed to be illiterate (which may apply only to English and not the ape language), and in episode #820: Space Mutiny, he inadvertently obstructed Pearl's attempts to free the "Mads" from captivity in Neronian Rome, and starts the Great Fire of Rome by knocking over a lamp in his haste to steal a wheel of cheese. In the final episode (#1013 *Danger: Diabolik*), he boasted of his new "job" at a zoo, possibly unaware of his future role as a specimen.

24.1 External links

- Professor Bobo at the Internet Movie Database

Chapter 25

RiffTrax

RiffTrax are comedy audio commentaries of movies, television programs and short films featuring *Mystery Science Theater 3000* (*MST3K*) comedians Michael J. Nelson, Kevin Murphy and Bill Corbett heckling (or riffing) films in the style of *MST3K*, a TV show in which Nelson was the head writer, and later the host. RiffTrax products are sold online and delivered by streaming video and download.

25.1 History

The site was launched by Nelson and Legend Films, now renamed Legend3D, in 2006 and is based in San Diego. In 2012, RiffTrax was purchased from Legend3D by Michael J. Nelson, Kevin Murphy, Bill Corbett and RiffTrax CEO David G. Martin. As of December 2014, RiffTrax had 13 employees.[1]

The main crew of Rifftrax (from left): Mike Nelson, Bill Corbett, and Kevin Murphy

The movies chosen for *Mystery Science Theater 3000* were predominantly low-budget B-movies because the show itself was low-budget and producers could only afford films with expired copyright or that had otherwise cheap licenses.[2] The idea of RiffTrax came about after *Mystery Science Theater 3000* was canceled and Nelson had researched and consulted a lawyer about the possibility of directly releasing DVDs of films with the commentaries included. But Nelson realized this initial idea was not feasible since he would be "sued out of existence."[3] Instead, the best way to distribute the commentaries would be to sell them independently of the films, to avoid having to obtain the rights to distribute the movies themselves.[4] There would be no legal or monetary restrictions to prevent Nelson from producing them,[5] though viewers would have to provide the movies themselves.

The early RiffTrax were almost all solo efforts on Mike's part, but it soon became apparent that there was a strong demand for them, and Nelson was quickly able to recruit more riffers for the project. Most official RiffTrax (not counting fan-made iRiffs and the spin-off RiffTrax Presents series) have a stable cast of Mike and former *Mystery Science Theater 3000* co-stars Kevin Murphy and Bill Corbett, a line-

up that happens to be identical to that of *The Film Crew* and the last three seasons of *MST3K*. That said, the guest slots vary often; other *MST3K* alumni have been featured, such as Mary Jo Pehl and Bridget Nelson, in addition to Internet personalities Richard Kyanka (of Something Awful fame), Josh Fruhlinger (writer of The Comics Curmudgeon) and Chad Vader, as well as actors Neil Patrick Harris, Fred Willard, and Joel McHale, and parodist "Weird Al" Yankovic. Nelson has said that he would like to bring in other guests.[6] The enthusiasm of guest riffers for the project led to the establishment of RiffTrax Presents, a series of tracks exclusively hosted by guest riffers and sanctioned by Nelson.[7] The success of the guest format is such that "Three Riffer Editions" of some films previously solo riffed by Mike have been produced for the VOD service, which feature new riffs by Mike in conjunction with Murphy and Corbett, and Mike has ceased producing solo riffs since 2007.

Along with the feature length tracks, Corbett, Murphy and Nelson have created riffs for a number of short films, typically educational and safety films, similar to the shorts presented before features on *Mystery Science Theater 3000*.

These include films by the Jam Handy Organization, Alfred P. Higgins Productions, Coronet Films and ACI Films, amongst others. Because these shorts are in the public domain, they can be downloaded with the commentary already recorded onto them. Shorts are usually released at least once, and often twice, a week

In 2008, RiffTrax launched iRiffs, which allows fans to upload commentaries to be sold on the website. iRiffs users are paid 50% of the net revenue generated by their products. iRiffs differentiates from normal RiffTrax in that both serious and humorous commentaries can be uploaded.[8] In February 2009, a contest was held by RiffTrax, in which a winning iRiffs user would be given $1,000 and a chance to develop a RiffTrax Presents title, receiving instruction and critique from Nelson, Murphy and Corbett.[9] The winners of the contest were Doug and Rob Walker and Brian Heinz of That Guy with the Glasses, who contributed an iRiff of *The Lion King*.[10] The RiffTrax commentary they produced was for *Batman Forever*.[11]

In October 2015, Rifftrax negotiated the rights to release available *Mystery Science Theater 3000* episodes through Vimeo via an all-access subscription plan, with a new rerelease uploaded each Monday; individual episodes could also be rented through the site, and in November 2015, RiffTrax also started to sell *Mystery Science Theater 3000* episodes on their website. The RiffTrax.com releases contain a newly recorded introduction on each episode by Michael J. Nelson, and a substantial percentage from the episode sales on the RiffTrax website goes to the cast members of MST3K.[12]

25.1.1   Live shows

Corbett, Nelson, and Murphy during the live Rifftrax of Birdemic

As part of SF Sketchfest in San Francisco, California, Nelson, Murphy, and Corbett have appeared several times performing live riffs alongside a screening of a film. As of 2008, they have appeared three times, having riffed

Daredevil and *Over the Top* in 2007 and *Plan 9 from Outer Space* in 2008, the last being shown in the historic Castro Theater.[13][14] The RiffTrax crew have done live internet broadcasts on Ustream.tv, riffing public domain films and taking viewer questions.

RiffTrax has teamed with NCM Fathom Events to host special one-night live RiffTrax events. Nelson, Corbett, and Murphy, along with special guests, perform their riffing to a live audience in one theater, which is simultaneously broadcast live to select theaters around the country, except in the Pacific Time zone, where a replay of the broadcast is shown. The first show was performed live at the Belcourt Theatre in Nashville, Tennessee on August 20, 2009, where Mike, Kevin, and Bill riffed *Plan 9 from Outer Space*, along with the short film *Flying Stewardess* and guest appearances by Veronica Belmont, Jonathan Coulton, and Rich Kyanka of Something Awful.[15] An encore showing was shown on October 8 of that year. The second RiffTrax Live! show in theaters took place on December 16, 2009, where they riffed several Christmas short films, one featuring special guest Weird Al Yankovic. An encore showing was shown in theaters the next day. The third RiffTrax Live! show took place on August 19, 2010, with the trio riffing the cult classic *Reefer Madness* (its encore showing took place on August 24).[16] A fourth live show riffing *House on Haunted Hill*, also filmed at the Belcourt,[17] occurred on October 28, 2010, with special guest Paul F. Tompkins. A fifth took place on August 17, 2011, featuring the film *Jack the Giant Killer*.[18] A sixth took place on August 16, 2012, featuring the film, *Manos, The Hands of Fate*.[19] A seventh live event took place on October 25, 2012, featuring *Birdemic: Shock and Terror*.[20] The live event featuring *Starship Troopers* took place on August 15, 2013.[21] The ninth live event featuring Night of the Living Dead was performed on October 24, 2013 .[22]

On February 25, 2013, RiffTrax announced a Kickstarter campaign to raise money to secure the rights to riff *Twilight*, the first film of *The Twilight Saga*, for their live show in August 2013.[23] Though the Kickstarter was successful, the RiffTrax could not work a deal to secure the rights to *Twilight* and that they had instead used the funding to secure *Starship Troopers*.[24]

On May 12, 2014, RiffTrax announced another Kickstarter campaign to raise money to secure the rights to the 1998 version of *Godzilla*.[25] The Kickstarter raised the goal of $100,000 within a day. On May 29, they announced a stretch goal of $250,000 to secure the rights to *Anaconda* for their Halloween show;[26] the goal was reached on June 10, the day before the Kickstarter's end.[27]

For 2015, RiffTrax repeated the Kickstarter approach, successfully funding its four-movie event for the year, dubbed "The Crappening". The successful Kickstarter will allow them to riff on *The Room*, *Sharknado 2: The Second One*, *Miami Connection*, and *Santa and the Ice Cream Bunny* at different live events throughout the year.[28] The Rifftrax riff of *The Room* will also be performed live during the 2015 Tribeca Film Festival's series of midnight special events.[29]

For 2016, RiffTrax used a Kickstarter campaign to fund, as of March 7, 2016, two live events with the possibility of two additional events, depending on how far they exceed their base goal. The first planned event is a riffing of the film *Time Chasers*, set for May 5, 2016; this event will use a high definition remaster of the film provided by its director, David Giancola. The second event is a *Mystery Science Theater 3000* cast member reunion, set for June 28, 2016, and will take place in Minneapolis, Minnesota. The format of the reunion will be similar to RiffTrax's SF Sketchfest "Night of the Shorts" shows, using various educational shorts as riffing material. The confirmed guests include Joel Hodgson, Bridget Nelson, Mary Jo Pehl, Trace Beaulieu, and Frank Conniff are confirmed alongside Nelson, Corbett, and Murphy, but other *MST3K* cast and crew may also participate, additionally, Jonah Ray, the new host for the revival of *MST3K* will also participate.[30][31][32] On March 11, 2016, their primary goal of $225,000 was reached.

(1) The name for this Fathom Event presentation was *RiffTrax Live: Christmas Shorts-Stravaganza* and is available for purchase under that name.

25.1.2 Television special

In conjunction with the National Geographic Channel, Nelson, Corbett, and Murphy was part of a 3-part television special, "Total Riff Off" that aired on the channel on April 1, 2014 and became available to buy as video-on-demand later on the RiffTrax site. Each part was one hour long featuring the three riffing on older National Geographic footage.[33] According to Nelson, the idea of the special came from a National Geographic producer who was also a fan of RiffTrax. RiffTrax and National Geographic worked together to find the best footage to work with for the special.[34]

25.2 The Rifftones

In August 2008, Nelson, Corbett and Murphy formed a musical trio named The Rifftones, initially to compete in Quick Stop Entertainment's second Masters of Song Fu competition. They won the competition, beating fellow musicians Jonathan Coulton, Paul and Storm and, in the final round, Jason Morris. After the competition, they decided to continue creating songs as The Rifftones, making songs based on the movies they've riffed and releasing a CD collecting

them, which is available on the RiffTrax site.

25.3 Use

RiffTrax commentaries are synchronized at the start of the movie using a cue.[35] To reassure consumers that the MP3 file is synchronized with the film, fictional character and riffer "DisembAudio" speaks occasional lines in exact synchronization with the movie.[36][37] "RiffTrax Presents" commentaries feature a female synchronization voice, Debbie.[38] Though RiffTrax are suggested to be played on an MP3 player or with computer software, they are sold as unrestricted MP3s, allowing users to choose the viewing method that suits them the best. A RiffTrax Player is also offered as a free download for Windows computers.[39]

The movies chosen for RiffTrax are based on two criteria: whether the movie lends itself towards a funny riffing, and whether the film is widely available on DVD.[35][37] These criteria have resulted in a wide variety of genre and era of movies chosen to be riffed. The first audio commentary made available through the web site in July 2006 was for the 1989 film *Road House*, long cited by Nelson as the cheesiest movie ever made.[40]

25.3.1 RiffTrax Player

The RiffTrax Player (RiffPlayer) is a program which automatically synchronizes the commentary playback to the DVD playback. The RiffTrax Player makes use of a commentary MP3 as well as a text file (.sync) containing the synchronization information of the DVD and the commentary. As of February 2015, the RiffTrax Player currently supports Microsoft Windows and Mac OS X Snow Leopard and up.

25.3.2 RiffTrax On Demand

RiffTrax On Demand features downloadable DRM-Free video files of films with RiffTrax commentaries embedded. RiffTrax On Demand has released many short, public domain, and educational films similar to the ones that *MST3K* would sometimes mock before a full-length movie began.[41]

25.4 Featured catalog

Main article: List of RiffTrax

25.5 References

[1] "Small Business Saturday!". *RiffTrax*. November 29, 2014. Retrieved April 27, 2015.

[2] Penny, Damian (2006-08-10). "Review: Rifftrax". Blog-critics Magazine.

[3] "Interview: Michael J. Nelson". TeeVee. 2006-10-16.

[4] Jurgensen, John (2006-11-18). "Everyone's a critic: DVD commentaries by fans". *Wall Street Journal*. p. P2. ISSN 0099-9660.

[5] Kaiser, Andy (2006-08-07). "Please talk during the movie". *The Grand Rapids Press*. p. D3.

[6] Salas, Randy (2006-08-11). "Cheeky remarks now on Riff-Trax". *Deseret News*. p. W06. ISSN 0745-4724.

[7] "RiffTrax Presents". RiffTrax. Retrieved 2008-08-20.

[8] "iRiffs FAQ". RiffTrax. Retrieved 18 November 2008.

[9] Legend Films, Inc. (February 4, 2009). "*RiffTrax Announces Chance for Fans to Work with the Stars of Mystery Science Theater 3000*". PR.com. Retrieved 28 February 2009.

[10] Walker, Doug (March 19, 2009). "Rifftrax Winner!". That Guy with the Glasses. Retrieved 2009-03-19.

[11] Walker, Doug (2 August 2009). RiffTrax: Batman Forever announcement. ThatGuyWiththeGlasses.com.

[12] Reich, J.E. (November 3, 2015). "RiffTrax Is Releasing Episodes of Cult Favorite 'Mystery Science Theater 3000'". Tech Times. Retrieved November 9, 2015.

[13] torgosPizza (May 20, 2007). "RiffTrax Live! for San Francisco Sketch Fest". RiffTrax. Retrieved 18 November 2008.

[14] Lastowka, Conor (December 11, 2007). "RiffTrax Live in San Francisco". RiffTrax. Retrieved 18 November 2008.

[15] Sampo (July 24, 2009). "Update: RiffTrax Live Event to Broadcast to Theaters". Satellite News. Retrieved 25 March 2011.

[16] "Mark your calendar: Next RiffTrax Live event scheduled for Aug. 19". mlive.com. Retrieved 15 June 2010.

[17] "The Stars of Mystery Science Theater 3000 Give the Vincent Price Classic House on Haunted Hill the RiffTrax Treatment on the Big Screen This Halloween". BusinessWire. October 5, 2010. Retrieved 25 March 2011.

[18] "RiffTrax LIVE: Jack the Giant Killer". Fathomevents.com. Retrieved 2011-08-15. with the short What Is Nothing and two short videos created by SomethingAwful.com

[19] Sloan, Will (16 August 2012). "'You Can't Just Be The Voice Of Generic Sarcasm": The Art of Movie Riffing". Retrieved 17 August 2012.

[20] "RiffTrax Live: BIRDEMIC". Fathom. Retrieved 17 August 2012.

[21] "RiffTrax Live: Starship Troopers". Fathom. Retrieved 15 August 2013.

[22] "RiffTrax Live: Night of the Living Dead". Fathom. Retrieved 29 October 2013.

[23] "RiffTrax Wants to Riff TWILIGHT Live in Theaters Nationwide! by RiffTrax: Mike, Bill & Kevin — Kickstarter". Retrieved 25 February 2013.

[24] "Major Announcement - We Have a Title!". *Kickstarter*. May 22, 2013. Retrieved April 27, 2015.

[25] "RiffTrax will Riff GODZILLA & ANACONDA in Cinemas Nationwide by RiffTrax: Mike, Bill & Kevin - Kickstarter". Retrieved 13 June 2013.

[26] "ANACONDA Stretch Goal!". Retrieved 13 June 2013.

[27] "ANACONDA is Happening!". Retrieved 13 June 2013.

[28] Jasper, Gavin (2015-02-19). "RiffTrax Live 2015 Schedule Announced: The Crappening". Den of Geek. Retrieved 2015-02-25.

[29] "HERE ARE THE FILMS IN 2015 TRIBECA FILM FESTIVAL MIDNIGHT SECTION". Tribeca Film Festival. 5 March 2015. Retrieved 5 March 2015.

[30] "RiffTrax Live 2016: MST3K Reunion, Time Chasers Live & More!". Kickstarter. 7 March 2016. Retrieved 7 March 2016.

[31] Hughes, William (March 9, 2016). "Get Involved, Internet: RiffTrax is putting together an MST3K reunion show". *A.V. Club*. Retrieved March 23, 2016.

[32] Rife, Katie (April 1, 2016). "Rifftrax Exclusive: Joel Hodgson and Jonah Ray join the live MST3K reunion". *A.V. Club*. Retrieved April 1, 2016.>

[33] Cooper, Gale Fashingbauer (2014-03-19). "No joke: Former 'Mystery Science' riffers get April Fool's Day special". The Today Show. Retrieved 2014-03-19.

[34] Stamato, Philip (2014-04-02). "Talking to Mike Nelson About RiffTrax's National Geographic TV Special". Splitsider. Retrieved 2014-04-02.

[35] "RiffTrax FAQ". RiffTrax. Archived from the original on 2008-01-21.

[36] "Mystery Science Theater, round two". The Commonwealth Times. 2007-02-08.

[37] "Mike Nelson Interview". Buzzgrinder. 2006-10-14.

[38] Murphy, Kevin; Corbett, Bill (May 13, 2008). *RiffTrax Presents: "Saw"*. RiffTrax.

[39] "RiffTrax Player". RiffTrax. Archived from the original on December 13, 2007. Retrieved 2007-11-10.

[40] "'Road House' -- it's the cheesiest, book says". *CNN.com*. August 16, 2000. Archived from the original on August 13, 2007. Retrieved April 28, 2015.

[41] "On Demand Catalogue". RiffTrax. Archived from the original on December 13, 2007. Retrieved 2007-12-21.

25.6   External links

- RiffTrax Official website

- @RiffTrax Official Twitter Account on Twitter

Chapter 26

Satellite of Love (Mystery Science Theater 3000)

The **Satellite of Love** (sometimes known as the **SOL**) is the fictional main setting of the comedy television series *Mystery Science Theater 3000*. It is a giant bone-shaped spacecraft that Joel Robinson (later replaced by Mike Nelson) and his friends — robots Crow, Tom Servo, Gypsy, Cambot, and the noncorporeal Magic Voice — live in. The vessel is in orbit above Earth during much of the series, except for a journey to the end of the universe [1] and its flight throughout the space-time continuum from Pearl Forrester. [2] Its name is a reference to the Lou Reed song, "Satellite of Love".

26.1   Story

According to the show's storyline, as part of an experiment to see how bad movies affected a person's mind, mad scientists Dr. Clayton Forrester and Dr. Laurence Erhardt (aka "The Mads") kidnap Joel, place him on the Satellite of Love, and shoot him into space. In order to keep himself from going mad, Joel builds his robot friends from parts of the spacecraft — namely the controls used to begin and end the movie. [3] Eventually, Gypsy incorrectly overhears that the Mads are finished with Joel, and plan to kill him, even though they really plan to fire their new temp, Mike Nelson. With the help of Mike, who discovers an escape pod in a box of "Hamdingers" after reading the SOL's manual, Gypsy forces Joel into it. Joel manages to safely return to Earth, where he crash-lands in the Australian Outback and eventually resumes his normal life. In response, the Mads kidnap Mike and send him up to the Satellite of Love to take Joel's place. [4]

At the end of season 10, the Satellite is deorbited and crash-lands near Milwaukee, Wisconsin. Mike and the Bots move into an apartment nearby, and are last seen starting to watch and riff *The Crawling Eye*, the movie which provided the first nationally broadcast episode of the series. Gypsy

founds an international conglomerate, ConGypsCo. [5]

In the unaired pilot, series creator Joel Hodgson (not yet using his character name "Joel Robinson") claims to have designed the Satellite of Love himself. [6] This was changed by the first aired episode to the more familiar storyline. The Satellite of Love during the KTMA era was also noticeably different from the version shown during the show's official run, built on a very low budget and not featuring the iconic "doggy-bone" shape. [7]

26.2   Layout

Little is known of the Satellite of Love's internal design and workings, following a general theme mentioned in the introductory song:

> *If you're wondering how he eats and breathes, and other science facts,*
>
> *Then repeat to yourself, "It's just a show, I should really just relax."* [8]

However, some aspects can be determined from the episodes. Based on the opening sequence of season 10, where the crew is briefly seen looking out a window, the room where the "host segments" take place is on the bridge, which appears to be in the upper-right sphere of the dog-bone-shaped ship. The crew is able to communicate with the Gizmonic Institute (KTMA era), Deep 13 (seasons 1-7), the apes' laboratory (early Season 8), Pearl's Microbus, "The Widowmaker" (late Season 8), and Castle Forrester (seasons 9-10) via a viewscreen of some sort (never shown on the TV series but shown in the movie), accessible through Cambot, which enables the crew to see the Mads and vice versa. (For all intents and purposes, it was the actual television camera that served this purpose, since whenever one group was communicating with the other, they looked at and spoke directly to the viewer.) Another viewscreen, the Hexfield, similarly provided two-way transmissions to spacecraft interiors and other locations. An exterior camera called Rocket Number Nine enabled the crew to see the entirety of the SOL and any spaceships or creatures in its immediate vicinity; the crew evidently viewed this through the same forward viewscreen that communicated with the Mads (again, the actual television camera).

Each of the SOL crew has his own living quarters, the locations of which are unclear. Mike's room was only seen once when the Bots were stalking him by putting a "hidden" camera on a toy. [9] Servo's room is very messy, with underwear strewn about (Servo collects underwear; how he managed to find all of it while having never been to Earth goes unexplained), and contains a car-shaped children's bed. [10] Crow's room has never been seen, nor has Gypsy's. It is unclear if Cambot had a room or if he simply lived on the bridge. From season 8 onward, the ship was also inhabited by billions of nanites, which apparently occupy the entire surface area of the ship's interior.

The Mystery Science Theater, in which the SOL captives watch and mock movies, is apparently located directly behind the bridge, on the opposite end of a long hallway with several oddly shaped doors (as suggested by the transition between host and movie segments), although the characters are rarely shown using it. As explained in one episode, life support throughout the rest of the ship is shut off while the movies are playing, thus forcing the crew to remain in the theater. [11] A few sightings include one instance of Mike and the Bots walking down the hallway, [12] and three occasions when someone runs into Cambot's direct passage before apparently being run over by Cambot himself — Joel and the Bots after a game of tag during the KTMA season, [13] repeated in a later nationally telecast episode, [14] and Mike alone after dislodging the stuck door in the season 10 opener. [15] Given the sheer size of the SOL, it seems unlikely that the hallway extended entirely from one end of the ship to the other, given that Joel and Mike were able to run from bridge to theater in so short a time; however, if it didn't, precisely what was at the other end of the SOL was never revealed. In Episode #410 (*Hercules Against the Moon Men*), Tom and Crow proved able to travel at least 50 yards (46 m) away from the theater doors while still on the non-theater side of the doors and corridor (i.e. at an angle to the viewer's left and forward, out of range of Cambot). However seeing as how Joel/Mike and the others always get to the bridge long before Cambot, it's possible their way into the theater is faster.

During the Joel era, Joel was sometimes seen jumping into a separate entrance beside the main door to get to the theater,[16] In later seasons this opening would be changed to, among other things, an airlock leading outside the satellite. [1] In a few episodes, this hatch allows the characters to send things back and forth between the satellite and Deep 13 via a long tube called the "Umbilicus", tethering the Satellite in orbit. In its first appearance, the Umbilicus is directly attached to Gypsy's snake-like body, and the characters can receive and send objects through Gypsy's mouth. Later, it is reconnected to an oven-like device on the SOL's bridge. [17] This tether is cut in the season 7 finale, causing the satellite to drift off into deep space. [1] It is implied that, during the later seasons, Brain Guy is responsible for keeping the Satellite of Love hovering above Earth with his telekinetic powers, especially with the last episode *Diabolik*, when Pearl douses his Brain with Mountain Dew and causes him to lose control of the SOL, thus sending it flying toward Earth.

The Satellite of Love is so large that many regions of it go unexplored, at least by Mike, who, during season 10, is surprised to learn that the SOL is equipped with a squash court, a silo, and a feed lot (to which Crow asks "Do you even live here, Mike?").[18] The presence of Joel's escape pod in "Docking Bay 14"[4] implies there are at least fourteen docking bays. The Satellite even has its own time machine which is built by the nanites during the episode *Terror from the Year 5000* (like the time machine from the movie, this machine looks suspiciously like a water heater). Unfortunately, due to its emitting dangerous radiation, only the bots (specifically Crow and Cambot) may use it, which they do a couple of times.

The ceiling to the bridge is impossibly high, as shown it can fit a flagpole in Merlin's Shop of Mystical Wonders and a mile-high pie in The Mole People, both of take Crow several minutes to fall off of. It also allows Tom and Crow to fly around in miniature planes in the episode The Skydivers. The bridge also has large air ducts, which Tom and Crow try to live in during Tormented. There are also rafters on the ceiling, which Professor Bobo and the Bot's pet chimpanzee Henry Kissinger hang from and threw microwaves from until Mike shot them down, all during Overdrawn At The Memory Bank. Crow also hides among these rafters in a host segment of 'Blood Waters of Dr Z', only to get himself stuck in them and Mike having to pry him out with a squeegee.

Another section, simply referred to as "the basement" by Tom Servo, is first featured in a scene where Crow attempts to "dig" his way back to Earth. [19] This part of the ship is located directly beneath the bridge. In season 8, Crow conducts an archaeological dig to uncover his forgotten 500-year solo occupation of the SOL since season 7. [20] The SOL crew also discovers albinos living in this area that happen to resemble the light-deprived Sumerian descendants of *The Mole People*. [21]

Also shown in the movie is a device referred to as the "manipulator arms", a set of robotic arms that extend from an opening near the back of the ship, controlled by a pair of virtual reality gloves. [12] The hands are labeled "Manos" (Spanish for "hands"), a joke referring to the infamous season 4 episode, *Manos: The Hands of Fate*. A similar mechanism appears in episode 104, *Women of the Prehistoric Planet*, but is never featured again.[22]

In Episode 820 "Space Mutiny," there are 3 escape ships on board the SOL, which Tom, Crow, and Gypsy pilot, but inevitably damages them beyond repair blasting at each other for fun and crashing them into the SOL. In that same episode, Tom Servo builds many unnecessary, inconvenient railings on the bridge, one around an unseen pit going straight down from the bridge. There was also a railing around Mike's seat in the theatre.

26.3   Footnotes

[1] *Mystery Science Theater 3000*, "Laserblast" [7.06].

[2] *MST3K*, "The Deadly Mantis" [8.04] through "The Projected Man" [9.01].

[3] *MST3K*, opening credits sequence, seasons K-5 (through "Mitchell" [5.12]).

[4] *MST3K*, "Mitchell" [5.12].

[5] *MST3K*, "Diabolik" [10.13].

[6] *The Mystery Science Theater 3000 Scrapbook*, host segment from unaired pilot "The Green Slime" [K.00]. VHS, Best Brains, Inc.

[7] *MST3K*, opening credits sequence, KTMA season ([K.01-K.22]).

[8] *MST3K* theme song, all episodes.

[9] *MST3K*, "Track of the Moon Beast" [10.07].

[10] *Mystery Science Theater 3000: The Movie* DVD (1996), ISBN 0-7832-1928-8 .

[11] *MST3K*, "Wild Rebels" [2.07]. Joel specifically refers to the viewing room as "the Mystery Science Theater" during the invention exchange.

[12] *MST3K: The Movie*, during the Hubble crash host segment.

[13] *MST3K*, "Humanoid Woman" [K.11]

[14] *MST3K*, "The Corpse Vanishes" [1.05]

[15] *MST3K*, "Soultaker" [10.01]

[16] *MST3K*, often shown in the opening credits.

[17] *MST3K*, "Girls Town" [6.01]

[18] *MST3K*, "Hamlet" [10.09]

[19] *MST3K: The Movie*, during the escape-attempt host segment.

[20] *MST3K*, "The Leech Woman" [8.02]

[21] *MST3K*, "The Mole People" [8.03]

[22] *MST3K*, "Women of the Prehistoric Planet" [1.04]

26.4   References

- *The Mystery Science Theater 3000 Amazing Colossal Episode Guide* (1996), ISBN 0-553-37783-3.

- *Satellite News: The Official Mystery Science Theater 3000 Web Site*

- *The Mystery Science Theater 3000 Scrapbook* VHS, Best Brains, Inc.

Chapter 27

The Film Crew

Box art for Hollywood After Dark

The Film Crew was a comedic team similar to *Mystery Science Theater 3000*, comprising former *MST3K* cast members Michael J. Nelson, Bill Corbett, and Kevin Murphy. They hosted Legend Films' colorized Three Stooges DVD release, packing in the four Stooge shorts that have fallen into the public domain: *Disorder in the Court* (1936), *Malice in the Palace* (1949), *Sing a Song of Six Pants*, and *Brideless Groom* (both 1947).

The Film Crew also maintained a website with humor columns and other content geared toward its fans.

27.1   On television

The Film Crew also occasionally hosted segments between movies on American Movie Classics, Sundance Channel, and the Starz/Encore cable channels in the United States.

In August 2005, during Encore's "Midnight Movies" schedule, The Film Crew provided introductions for the documentary on the subject *Midnight Movies: From the Margin to the Mainstream* and for the cult classics *The Rocky Horror Picture Show*, *Reefer Madness*, *Night of the Living Dead*, *The Harder They Come*, and *Pink Flamingos*.

27.2   DVD Riffs

On October 19, 2006, it was announced that The Film Crew would be providing commentary tracks for a series of B-movies. To promote their announcement, the Crew offered a poll on their website asking viewers to sample clips of each film and choose which they would prefer to see released first out of the four titles announced.

The four titles, in order of votes received (and, as a result, their release order), are:

- *Hollywood After Dark*, starring Rue McClanahan (of *The Golden Girls* fame)

- *Killers from Space*, starring Peter Graves

- *The Wild Women of Wongo*

- *The Giant of Marathon*, starring Steve Reeves

The episodes were produced in association with Rhino Entertainment, which was to distribute the episodes on DVD. However, Jim Mallon of Best Brains approached Rhino, and threatened to pull future releases of MST3K from Rhino's distribution unless they passed on The Film Crew (Mallon claimed that it was "too similar to MST3K", forcing Rhino to choose between MST3K or Film Crew).

Rhino then ended its relationship with The Film Crew. It wasn't until 2007 that arrangements were made with Shout! Factory to release the material (requiring some "looping" of lines from the original scripts – specifically, "Bob Honcho" was originally named "Bob Rhino," and this had to be changed due to Rhino no longer being the distributor).

Releases were in NTSC format but have no region encoding. Each released episode contains at least one "extra" ("Ode to Lunch" in *Hollywood After Dark* and "Did You Know..." in *Killers from Space*).

27.2.1   Premise

The Film Crew, stuck in the confines of a basement at work, lays down commentary tracks to every obscure movie dished to them by their boss, Bob Honcho. As part of their job, each of the three wears a matching "working-class" uniform and an unwieldy headset while riffing on each film. Each release contained a short "Lunch Break" sketch, in which they would act out a humorous sketch.

27.2.2   End

During the delay when there was no distributor for the Film Crew DVDs, the sets were destroyed and the cast moved on to the similar project RiffTrax, complicating the issue of any possible future Film Crew episodes.[1] Kevin Murphy has indicated that "We haven't gotten any new orders from Shout Factory – the new home of MST by the way – and since RiffTrax is becoming so much fun, I think you may have seen the last of the Film Crew."[2]

As of July 13, 2008, The Film Crew's former website, filmcrewonline.com, is no longer online. Shortly thereafter, Shout Factory put three Film Crew movies on Hulu.com. In April 5, 2009, all four movies became available on YouTube.

However, on February 5, 2016, Rifftrax began offering the Film Crew episodes for sale, starting with *Killers From Space*.[3]

27.3   References

[1] Corbett, Bill (December 20, 2007). "Talk to me, peoples". RiffTrax. Archived from the original (blog) on 2008-10-17. Retrieved 2012-11-01.

[2] Murphy, Kevin (February 20, 2008). "I Got Yer Twenty Questions, Right Here!". RiffTrax. Archived from the original (blog) on 2009-04-22. Retrieved 2012-11-01.

[3] http://www.rifftrax.com/the-film-crew-killers-from-space

27.4   External links

- The Film Crew on RiffTrax
- The Film Crew at Shout! Factory
- The Film Crew on Hulu
- *The Film Crew: Hollywood After Dark* at the Internet Movie Database
- *The Film Crew: Killers from Space* at the Internet Movie Database
- *The Film Crew: Wild Women of Wongo* at the Internet Movie Database
- *The Film Crew: The Giant of Marathon* at the Internet Movie Database

Chapter 28

Timmy Big Hands

Timmy Big Hands was a humor webzine. Created in 2000 by the former cast and crew of *Mystery Science Theater 3000*, the site garnered much critical acclaim and accolades, but eventually was retired.

The site featured odd but humorous reviews of everyday items, comics, strange games, and new syrup ads each week. [1]

The games included such oddities as "Kill-a-Guy" where you would literally play God and kill a man by simply clicking on him, as well as an interactive game called "Apologize to Steve", in which the concept was to apologize to Steve.

Comics included "The Cliparts" which were simply crafted from clip art and given dialogue balloons. The stories were usually nonsensical and the art would run the gamut from office workers to suddenly having a tall Indian enter the frame.

In 2001, the site was put up for sale on eBay. The new owners decided not to continue with the project.

In an interview in 2003, Mike Nelson stated:

28.1 Contributors

- Michael J. Nelson
- Kevin Murphy
- Bill Corbett
- Patrick Brantseg
- Paul Chaplin

28.2 References

[1] TimmyBigHands

[2] Paul Czarnowski. "Michael J. Nelson". *CRC Pulse*. CRC Radio. Archived from the original on 2007-03-10.

28.3 External links

- Jennifer Owens (2000-05-01). "MST3K Guys' Big-Hand Job". Brandweek.

Chapter 29

Tom Servo

Tom Servo is a fictional character from the American science fiction comedy television show *Mystery Science Theater 3000* (*MST3K*). Tom is one of two wise-cracking, robotic main characters of the show, built by Joel Robinson to act as a companion and help stave off madness as he was forced to watch low-quality films (ironically, he, along with the other bots, is made from the parts that would have otherwise allowed Joel to actually control the film). At least during the Comedy Central era, he was somewhat more mature and cynical than his companion Crow T. Robot. Tom, more often than the others, signals the need to exit the theater to perform host segments.

29.1   Overview

Tom Servo is a red puppet that has a gumball machine (Carousel Executive Snack Dispenser) for a head, a body composed of a toy "Money Lover Barrel" coin bank and a toy car engine block, and a bowl-shaped hovercraft skirt (a Halloween "Boo Bowl") instead of legs. His arms are a pair of small white ventriloquist's dummy hands on the ends of springs that are not really functional as arms, a point that is commented on occasionally throughout the series. Some episodes feature Tom with objects already in his hands, raising the unanswered question of how they got there. His shoulders are made from the front of an Eveready Floating Lantern. Because Servo's head is transparent, chromakeyed images appear projected through it, and thus a second puppet was built for use in the theater segments, entirely spray-painted black. This black Servo also appeared in a host segment in episode #609, *The Skydivers*.

Servo's appearance has changed over time. In the pilot for *MST3K*, the robot who would become Servo was named "Beeper," who just spoke in beeps that only Crow could understand (similar to R2-D2 and C-3PO from the *Star Wars* films). He was an all-silver robot vaguely shaped like the ultimate Servo, with funnel-shaped shoulders, silver rubber tube arms, a plastic flowerpot for a hoverskirt, and a small fishbowl for a head. He was renamed 'Servo' after a vending

machine called the Servotron. Sometime between the pilot and episode #K03: *Star Force: Fugitive Alien II* Servo's head was replaced with the now-familiar gumball machine for most of the series. (*MST3K* archival site mst3ktemple.com presents a substantive argument that this replacement was probably done before or during episode #K01: *Invaders from the Deep*, based on an analysis of related set and costume changes.).

In Season 1 on the Comedy Channel, he was given a red color, longer black tube arms, squared white shoulders, a different hoverskirt, and the Carousel Snack Dispenser gumball machine head with a white beak. Around episode 105: *The Corpse Vanishes*, Servo's head was replaced with a slightly modified version of his "Carousel" head. The "neck" was slightly wider and the beak (now silver at this point) appeared smaller. This version of Servo's head would be used for the remainder of Season 1. For Season 2, the black tubing used for his arms was replaced by a pair of small silver springs and the more familiar Carousel Dispenser head design (KTMA/pre episode 105) returned with a silver beak. This physical form was pretty much the same throughout the remainder of the series, save for a brief flirtation (during episodes #205: Rocket Attack USA and #206: Ring of Terror) with a slim cylindrical gumball-machine head to try to reduce the screen area Servo's head obscured. It was introduced as a "haircut" that Joel gave Servo, but was quickly abandoned. By mid season 3 an extra cap from another Carousel Dispenser was added just below the "bubble" making Tom's head appear slightly taller and slimmer. Briefly in early season 4, Servo's white hands were changed to beige before returning to white after only a few episodes. During a host segment in the Sci-Fi era, he briefly acquired the body of a "beautiful butterfly."

Servo's voice and personality also changed during the show's early years. While Josh Weinstein operated Servo during the KTMA season, Servo originally spoke with a sleasy, Bullwinkle-type voice in episode K02 (the earliest seen of the regular Servo puppet), then a rather slowly with a squeaky voice from K03-K05, and was somewhat immobile during host segments but oddly very active in the the-

Kevin Murphy, pictured in 2012, who operated and voiced Servo after Josh Weinstein

ater. In episode K06, Weinstein switched to a lower voice that Servo repeatedly proclaimed as his new "MIGHTY VOICE!" When Weinstein left at the end of Season 1, Kevin Murphy took over Servo's operation and tried to match Weinstein's Servo voice and personality, but gradually developed a somewhat new Servo sound and character (though Murphy has a fairly deep voice himself). This was explained as tinkering by Joel. During Murphy's tenure, Servo took many opportunities to showcase his excellent singing. He also has an extensive underwear collection (as seen in *Mystery Science Theater 3000: The Movie*), as well as a large number of duplicates of himself that he made in episode #420: *The Human Duplicators* (also seen in episode #612: *The Starfighters*, episode #910: *The Touch of Satan*, episode #913: *Quest of the Delta Knights*, episode #1003: *Merlin's Shop of Mystical Wonders*, episode #1004: *Future War*, and episode #1013: *Diabolik*). Female Servo duplicates were featured in The Touch of Satan and Quest of the Delta Knights.

Whenever a member of the cast is required to dress in drag for a sketch, Servo usually does the honors. This is both because of the dichotomy of women's clothes amusingly

contrasted with puppeteer Murphy's strong baritone voice and because, in Murphy's words, "Servo looks better in a dress than Crow." Also, Servo is the only robot (other than Cambot in seasons 5-10) whose entire body can seen on the show, since Crow's legs are behind the desk and Gypsy's body is several yards long.

Servo normally has a condescending personality and at times can make literary and technical references that are above his companion's heads. He frequently attempts to seem physically imposing to others, once acquiring "lifts" for his hover skirt to increase his size (accused by Mike of suffering "short man's disease") and on another occasion showing off a small arsenal he had acquired while drifting through space. Almost invariably, however, any attempts at confronting danger or displaying his intellectual skill cause him such frustration that he ends up crying, often needing consolation from Joel or Mike.

Furthermore, he's easily rattled by sarcastic remarks made from Crow, such as the time when he made fun of his infatuation with a boy's pet turtle Tibby in the episode 'Gamera'.

He does have a good understanding and intellect in spite of his sensitivity and frustration, and has revealed in the episode 'The Gunslinger' that he's able to teleport at will, even though he only demonstrates this on rare other occasions.

29.2 Other appearances

- Tom Servo also appeared in the *Cops*-style *Star Wars* spoof *Troops* as a droid purloined by Jawas. *Star Trek: Deep Space Nine* is also known to contain a reference to Tom Servo's Used Robots in the directory to the station's Promenade business area, though it is unlikely that this reference ever appeared onscreen.

- Servo, along with Crow, has a cameo appearance (appropriately in silhouette) in the Futurama episode "Raging Bender". Ironically, he shushes the main characters as they begin to riff a newsreel they were watching.

- A robot bearing at least a striking resemblance to Tom Servo appears in the Beck music video "Where It's At."

- Servo makes an appearance in silhouette in the Homestar Runner cartoon "A Jorb Well Done", during a short scene in a theater.

- Though not actually Tom Servo, a stolen Gumball machine featured in the Hoax II, skate video is lovingly named Tom Servo. He unfortunately gets smashed by accident a short while after being purloined.

- Tom (with his cylinder head) and Crow are seen in a Star Wars Tales comic about a Force using droid. He is also seen, along with Crow and Gypsy, in Tag and Bink: Revenge of the Clone Menace.

- In the Archie Comics series Sonic the Hedgehog Issue#52, Sonic does battle with three robots, resembling Tom Servo, Crow, and Cambot.

- The prototype web browser Servo is named after Tom Servo

- In Gold Digger/Ninja High School "a science affair" issue 1 Tom servo can be seen in the background. Marked as "Servo-tron"

29.3   References

- *The Mystery Science Theater 3000 Amazing Colossal Episode Guide* (1996), ISBN 0-553-37783-3.

- *The Official Mystery Science Theater 3000 Bot Building Booklet* (1998), Best Brains, Inc., ISBFE 05557143431.

- *Satellite News: The Official Mystery Science Theater 3000 Web Site*

- *The Mystery Science Theater 3000 Scrapbook* VHS, Best Brains, Inc.

- *Mystery Science Theater 3000: The Movie* DVD (1996).

- *Mystery Science Theater 3000*, episode #K02 (*Revenge of the Mysterons*)

- *Mystery Science Theater 3000*, episode #K06 (*Gamera vs. Gaos*)

- *Mystery Science Theater 3000*, episode #201 (*Rocketship X-M*)

- *Mystery Science Theater 3000*, episode #205 (*Rocket Attack U.S.A.*)

- *Mystery Science Theater 3000*, episode #206 (*Ring of Terror*)

- *Mystery Science Theater 3000*, episode #1004 (*Future War*)

- *Mystery Science Theater 3000*, episode #1013 (*Diabolik*)

29.4   External links

- A page with instructions for building a Tom Servo

- Parts list for the above link

- A guide to the different heads used on the Tom Servo puppet

- Details of Tom Servo's construction through the entire run of the series.

Chapter 30

TV's Frank

TV's Frank, played by Frank Conniff, is a fictional character, mad scientist Dr. Clayton Forrester's lab assistant in the television comedy series *Mystery Science Theater 3000*. He appears at the beginning of Season 2, with the departure of Forrester's earlier co-scientist Dr. Laurence Erhardt, and continues through Season 6. According to *The Mystery Science Theater 3000 Amazing Colossal Episode Guide*, Dr. Forrester discovered Frank working at a nearby Arby's. Early on he was simply called Frank; later he acquired the more ostentatious name which is a reference to how a TV personality would sometimes be introduced as "TV's so-and-so" on talk shows and other programming.[1] Frank wears a black chauffeur's uniform and his hairstyle includes a spit-curl (resulting in a resemblance to that of Marlon Brando's Jor-El role in the 1978 *Superman* film). He has an unusual habit of calling Dr. Forrester "Steve". He is listed in Deep 13's employee records as "Frank, TV's," indicating that "Frank" is actually his surname and "TV's" is his given name.

Little is known about Frank's past save that he attended Harriet Tubman High School (a real high school in Compton, California[2]), where he was held back at least twice. While working at Arby's, Frank was allegedly nicknamed "Zeppo" due to his supposed sense of humor, which was somewhat ironic since Zeppo was the least funny Marx Brother. Frank had a surprisingly large personal fortune which surfaced whenever a large amount of money was required for a particular skit.

30.1   Tenure

TV's Frank's first *MST3K* episode was episode #201 *Rocketship X-M*, where, apparently still in his Arby's mindset, he took fast-food orders and, rather to Dr. Forrester's annoyance, almost brought the Satellite of Love down so the crew could "dine in." His last regular appearance was episode #624 *Samson vs. the Vampire Women*, during which he was assumed into "Second-Banana Heaven", where sidekicks and henchmen could live in a peaceful paradise without fear of reprisal from their cruel masters, by the angel Torgo the White. Dr. Forrester was actually very saddened when Frank left him for *Second Banana Heaven*, even lamenting his loss in a song entitled *"Who Will I Kill?"*. Frank later appeared to the despondent Forrester as an otherworldly entity and "reconciled" with him, even agreeing to "push the button" one last time. After his departure, he was immortalized in the following year's *Mystery Science Theater 3000: The Movie* as a door handle (Door 2) on the way to the theater.

Frank also made a guest appearance in the Season 10 opener *Soultaker*, having gotten a job in the afterlife as a Soultaker after complaining that Second-Banana Heaven was "too political" and that Pat Buttram "had it in" for him. (This episode also features a cameo by Joel Hodgson, the show's creator, who played SOL resident Joel Robinson for the show's first six years.) In this appearance, he took the soul of Bobo and played ring toss with it in Castle Forrester. He is also mentioned in episode #822: *Overdrawn at the Memory Bank*; as the Novacorp chairman, who bears a resemblance to a grossly overweight Frank, appears during the opening credits, Mike says, "Wow! TV's Frank! Hi, Frank!" Servo adding, "He's come up in the world!"

During TV's Frank's tenure, the catchphrase "Push the button, Frank!" was a staple of most shows and the last thing heard before the credits would roll.

30.2   Role

Frank usually serves as a foil to his evil boss, Dr. Forrester, and is frequently on the receiving end of many of Forrester's experiments or punishments. He is subjected to many painful deaths but always returns alive and well shortly afterwards; whether this rapid recovery is a quality bestowed upon him by Dr. Forrester or a talent that Frank always possessed (making him uniquely qualified as a mad scientist's guinea pig) was never explained, since it was, after all, "just a show." In the episode *Laserblast*, Pearl finds his spare head in a box, Dr. Forrester having made

Frank's head explode years earlier in *Gunslinger*. Occasionally, however, Frank gets revenge on his taskmaster by directly or indirectly causing Forrester harm. One significant example of sidekick payback occurs in episode #619 *Red Zone Cuba*, when Frank, supposedly owing the mob "50 large", passes Forrester off as himself, earning the evil scientist two severe beatings and an episode-long stint in full-body bandages.

Frank participates in the weekly invention exchanges that are primarily a feature of the Joel Robinson years. His first invention was a rip-off of Joel's invention, the BGC-1.9 drum machine. He repeated his thievery by introducing the "Cheese Phone", which Joel had supposedly recorded in his notes from the 1970s. Unlike Forrester (who rarely even remembered Crow and Servo's names), Frank took a liking to Joel (and later Mike) and the 'Bots, who reciprocated his friendship.

In Season 6, Dr. Forrester discovers that his mother, Mrs. Pearl Forrester, has had a strong friendship with TV's Frank that he'd been unaware of. When she comes to visit, she winds up spending all her time with Frank and neglecting her son, suggesting some of the formative influences that made Dr. Forrester an evil scientist.

30.3   Behind the scenes

Frank Conniff was one of the *MST3K* writers along with Tory Kicinski, providing his share of the movie "riffs" that defined the show. He also frequently screened the movies that ended up as *MST3K* fodder. Occasionally, Frank's multiple roles had an impact on the show's storyline. In episode #621 *The Beast of Yucca Flats*, one host segment sketch features Crow T. Robot continually asking if it's 11:30 yet, based on breakfast-skipping Conniff's obsession with having lunch at 11:30.

On two occasions, the Mads try their hand at riffing. In episode #323: *The Castle of Fu Manchu*, taunted by Joel and the Bots, they make a weak attempt at riffing, and fail miserably. In episode #611: *Last of the Wild Horses*, in a parody of *Star Trek* episode "Mirror, Mirror", TV's Frank and Dr. Forrester become the SOL captives and spend part of the episode providing all the riffs. Dr. Forrester comments as they enter the theater that he wasn't going to carry Frank in, a reference to Joel or Mike regularly carrying Tom Servo into the theater.

30.4   In Popular Culture

In several episodes of the Nickelodeon series *Invader Zim* Tv's Frank can be seen in random scenes. On the DVD

commentary for these episodes it is confirmed that it was intentional as Frank Conniff also wrote for *Invader Zim*.

30.5   References

[1] MST3K FAQ - Where are they now?

[2] http://www.greatschools.net/modperl/browse_school/ca/1616

- Beaulieu, Trace, et al. *The Mystery Science Theater 3000 Amazing Colossal Episode Guide* (1996). ISBN 0-553-37783-3.

- The Official *MST3K* FAQ

30.6   External links

- TV's Frank at the Internet Movie Database

Chapter 31

WUCW

WUCW is the CW-affiliated television station serving Minneapolis and Saint Paul, Minnesota. It broadcasts a high definition digital signal on UHF channel 22 (virtual channel 23.1 via PSIP) from a transmitter at the Telefarm site in Shoreview. Owned by the Sinclair Broadcast Group, WUCW maintains studios on Como Avenue (near the Minnesota State Fairgrounds) in Saint Paul. Syndicated programming on this station includes *Extra*, *The People's Court*, *The Steve Wilkos Show*, and *Family Feud*. The station is perhaps best known for originating the cult cable television series *Mystery Science Theater 3000*, which got its start as a locally produced program when channel 23 was an independent station.

31.1 History

31.1.1 As an independent station

KTMA logo, recognized worldwide by fans of Mystery Science Theater 3000 *from videotapes of KTMA-era episodes, used from 1986 to 1992.*

Channel 23 signed on the air on September 22, 1982, under the callsign **KTMA** (the call sign standing for "**T**win **M**etro **A**rea"), as the Twin Cities affiliate of the Spectrum over-the-air subscription television service. In addition to mainstream and softcore pornographic films, and animated fare such as *Grendizer* from Spectrum, the station also broadcast home games of the MLB's Minnesota Twins and the

NHL's Minnesota North Stars (now the Dallas Stars). The Spectrum programming lasted on the station for just two years, before the station was sold to the United Cable TV Corporation, who in turn began asking for bids one year later in 1985. The owners of radio station KTWN made a bid to purchase channel 23 and took some operational control of the station for a while, broadcasting music videos. However, the station was eventually sold to the KTMA-TV Acquisition Corp. in 1986 for $7 million. The station's new general manager Donald W. O'Connor soon transitioned KTMA to a more traditional general entertainment station, acquiring a number of older syndicated programs such as *The Andy Griffith Show* and *Laurel and Hardy*.

Despite a major marketing campaign in 1987, after the station was acquired, the station was only moderately successful at attracting viewers and revenue from commercial advertising. In 1988, attempts were made at creating locally produced shows. To fill a hole in the Saturday night lineup, the station created "Saturday Night at Ringside". A multi-hour block of pro wrestling programming hosted by Mick Karch which debuted in March 1988 and lasted until 1992.

As production manager Jim Mallon sought to fill a hole in the Sunday night lineup, he talked to his contacts in the local comedy community and ended up meeting Joel Hodgson. After a successful lunch meeting with Mallon to produce a new locally produced program for KTMA, Hodgson created *Mystery Science Theater 3000* (also known under the abbreviated title *MST3K*), which debuted in November 1988.

In December of that year, KTMA attempted to create a new regional television network called the *Minnesota Independent Network* (MIN), in conjunction with a media group based in Fargo, North Dakota (KVRR channel 15), KXLI (channel 41) in St. Cloud and KXLT-TV (channel 47) in Rochester. Despite good intentions, the network never got off the ground. KVRR, the Fargo area's Fox affiliate, continued its normal operations, while KXLI eventually was forced to go off the air for two years. KTMA was also hit hard, leading O'Connor to file for bankruptcy reorganization in July 1989. Hodgson and Mallon sold *Mystery Sci-*

ence Theater 3000 to cable network, The Comedy Channel (now Comedy Central) that year; the program ran on the network for seven years, before moving to the Sci Fi Channel for its final three seasons. In 1989, a small start-up home shopping network called Valuevision (now ShopHQ) rented space at the channel 23 studios and made its initial launch on two area low-power stations broadcasting on VHF channel 7 and UHF channel 62.

Through the bankruptcy, the station still maintained its general entertainment programming format, partially supported with infomercials, paid religious shows, daytime shows from the major broadcast networks that KARE, KSTP-TV and WCCO-TV did not clear for broadcast (mostly game shows from CBS and ABC) and home shopping programming. In December 1989, KTMA moved into a studio facility near the Minnesota State Fairgrounds in Saint Paul that was formerly occupied by PBS member station KTCA (that station had already moved into a newly constructed studio building in downtown Saint Paul). After nearly two more years of bankruptcy proceedings, O'Connor was fired as general manager by the court-appointed trustee. In November 1991, the station was purchased by Christian broadcaster Lakeland Group Television.

Under Lakeland Group ownership, channel 23 adopted a family-oriented programming format, and changed the station's callsign to **KLGT** (standing for either "light" or "Lakeland Group Television") in 1992, using the on-air branding "Sonlight 23". Programming during this period consisted of a few hours of religious programming a day, along with family-oriented off-network sitcoms, cartoons and classic movies. The new format was not very popular, but the station held its own. Lakeland Group brought sports programming back to the station in 1994, this time in the form of the St. Paul Saints minor league baseball, basketball games from the NBA's Minnesota Timberwolves, and games from the Minnesota Moose minor league hockey team. Around this time, KLGT began an association with CBS affiliate WCCO-TV, with that station providing local news updates during KLGT's prime time programming.

31.1.2   Joining The WB

In January 1995, KLGT became the Twin Cities' charter affiliate of the fledgling WB Television Network. KLGT did not air a news program of its own until WCCO purchased airtime on the station in order to air an experimental newscast known as *News of Your Choice* in 1995, in which two different newscasts (each one covering different stories) were produced simultaneously at the WCCO studios. At regular intervals, the news anchors would mention the stories airing on either station, allowing viewers to decide

KLGT "WB23" logo, used from 1995 to 1998.

which one they were more interested in and to tune into the appropriate station. Due to declining ratings at WCCO at the time, the project was canceled in January 1996 after one year.

Following the station's 1998 sale to the Sinclair Broadcast Group, its callsign was again changed to KMWB (standing for "Minnesota's WB"). The station was hit by the 2004 controversy surrounding the decision by Sinclair to air the documentary *Stolen Honor*, which was critical of U.S. presidential candidate John Kerry's service record in the Vietnam War.

31.1.3   From The WB to The CW

Logo as "The CW Twin Cities," from 2006 to 2013.

On January 24, 2006, CBS Corporation and the Warner Bros. unit of Time Warner made the decision to merge UPN and The WB to form a new network called The CW Television Network.[1] On February 22, News Corporation announced a new competing network, MyNetworkTV.[2]

On May 2, 2006, Sinclair Broadcast Group signed an affiliation agreement with The CW for the company's eight WB affiliates to join the network; as a result, KMWB was confirmed as the new network's Twin Cities affiliate.[3] UPN affiliate WFTC would automatically affiliate with MyNetworkTV, due to the station being owned by that network's owner, the News Corporation subsidiary Fox Television Stations.

In preparation for the affiliation switch, KMWB changed its call sign to **WUCW** (in reference to The CW and predecessors The WB and UPN) on June 19, 2006, to reflect its forthcoming CW affiliation. Ironically despite three call letter changes, WUCW's licensee is still listed as "KLGT Licensee." On August 16, 2006, WUCW changed its on-air branding from "WB23" to "CW Twin Cities" and then in 2013 to "CW23". Until recently, the station carried Minnesota Timberwolves basketball games. WUCW continues to support homegrown comedy, airing the first *Transylvania Television* special on October 12, 2007, and the new one-hour *TVTV Halloween Special* on October 22, 2010. And in 2014, this station revived a cult favorite from the '80s, the "Melon Drop" New Year's Eve TV special, and followed it up in 2015.

31.2 Digital television

31.2.1 Digital channels

The station's digital channel is multiplexed:

On July 15, 2006, WUCW launched a second digital subchannel, which carried programming from The Tube Music Network; it was part of an affiliation agreement between The Tube and Sinclair, the latter of whom severed the agreement at the beginning of 2007 (The Tube would discontinue operations that October). Subchannel 23.2 remained silent until joining another music video network, TheCoolTV, on October 4, 2010. After Sinclair's affiliation with TheCoolTV concluded at the end of August 2012, 23.2 returned to silent status until joining classic movie network GetTV in July 2014.

On October 22, 2010, WUCW launched subchannel 23.3, which carried music videos and programming from The Country Network, which was renamed ZUUS Country in Summer 2013. ZUUS Country would be replaced on 23.3 with the male-oriented network Grit in December 2014.

31.2.2 Analog-to-digital conversion

WUCW (as with most Sinclair-owned television stations) shut down its analog signal, over UHF channel 23, on February 17, 2009, the original target date in which full-power television stations in the United States were to transition from analog to digital broadcasts under federal mandate (which was later pushed back to June 12, 2009). The station's digital signal remained on its pre-transition UHF channel 22.[5][6][7][8][9] Through the use of PSIP, digital television receivers display the station's virtual channel as its former UHF analog channel 23.

31.3 Translators

The broadcast signal of WUCW is also extended by way of six digital translators in central and southern Minnesota:

List of translators

31.4 References

[1] 'Gilmore Girls' meet 'Smackdown'; CW Network to combine WB, UPN in CBS-Warner venture beginning in September, CNNMoney.com, January 24, 2006.

[2] News Corp. Unveils My Network TV, *Broadcasting & Cable*, February 22, 2006.

[3] Eight Sinclair Stations Sign On With CW, *Broadcasting & Cable*, May 2, 2006.

[4] RabbitEars TV Query for WUCW

[5] http://www.jamestownsun.com/ap/index.cfm?page=view&id=D9667KUO3

[6] http://wcco.com/digital/digital.switch.tv.2.928751.html

[7] Local TV station goes digital-only, Minnesota Public Radio, February 18, 2009

[8] "DTV Tentative Channel Designations for the First and the Second Rounds" (PDF). Retrieved 2012-03-24.

[9] CDBS Print

- "TV23 History" (archived file downloaded October 2003 or earlier) from NebraskaRadio.com

31.5 External links

- TheCW23.com - Official website

- RabbitEars.info query of WUCW

- Query the FCC's TV station database for WUCW

- Query the FCC's TV station database for K14LF-D
- Query the FCC's TV station database for K23FY-D
- Query the FCC's TV station database for K26CL-D
- Query the FCC's TV station database for K32GX-D
- Query the FCC's TV station database for K38LC-D
- Query the FCC's TV station database for K43MJ-D

- BIAfn's Media Web Database -- Information on WUCW-TV

31.6 Text and image sources, contributors, and licenses

31.6.1 Text

- **Best Brains** *Source:* https://en.wikipedia.org/wiki/Best_Brains?oldid=694271244 *Contributors:* WhisperToMe, Jeffq, Rich Farmbrough, Brian0918, Szyslak, Robert K S, JBellis, Koavf, CR85747, Nihiltres, Rob T Firefly, Rubber cat, CambridgeBayWeather, Bkberry, EmiOf-Brie, Pegship, SmackBot, Hmains, Sabrewing, Alaibot, Slysplace, Duvallg, Yobot, AnomieBOT, DoctorJoeE, BurgerKingFanatic, 19jduryea, MindedDionysus, Nsteffel and Anonymous: 9

- **Cambot** *Source:* https://en.wikipedia.org/wiki/Cambot?oldid=693397279 *Contributors:* Jeffq, Gamaliel, Mike R, Jesster79, MakeRocket-GoNow, Thanos6, Ashley Pomeroy, Smoke, CapnJim, Wack'd, Rob T Firefly, Misza13, EmiOfBrie, Betacommand, Andy120290, BNLfan53, Leechcode5, Ninski, Sabrewing, Cydebot, Bantab, JAF1970, Tom servo, RobotG, EagleFan, Weirdman, Rwe1138, RJASE1, Slysplace, Bot-Multichill, Green-eyed girl, ImageRemovalBot, Tyrantulas, Polly, Addbot, Fraggle81, FrescoBot, JIK1975, Haon 2.0, GoingBatty, Evan-Amos, IJVin, PhnomPencil, Mtbrunson, Godzilla and mst3k and Anonymous: 32

- **Cartoon Dump** *Source:* https://en.wikipedia.org/wiki/Cartoon_Dump?oldid=724360488 *Contributors:* Rich Farmbrough, Ashley Pomeroy, Koavf, Wack'd, Sugar Bear, SmackBot, GrizzlyFlats, Alaibot, JamesAM, Lightmouse, Yonskii, Yobot, Krelnik, Jj98 and Anonymous: 5

- **Cinematic Titanic** *Source:* https://en.wikipedia.org/wiki/Cinematic_Titanic?oldid=718376589 *Contributors:* Jeffq, D6, Druid816, Redfarmer, Erik, Robert K S, Rjwilmsi, Koavf, Sdornan, Avalyn, Wowbobwow12, Jason.cinema, Wack'd, Rob T Firefly, Ericorbit, Nikkimaria, PortnoySLP, JQF, NielsenGW, Sugar Bear, Bdve, SmackBot, McGeddon, Mattymatt, Jlahorn, Jkp1187, Rolypolyman, Toughpigs, Lostinlodos, Andy120290, Ithizar, Shamrox, Thornghost, Vjamesv, Jere7my, CmdrObot, ShelfSkewed, Ken Gallager, Alaibot, Thijs!bot, Maddyfan, Farosdaughter, Sree-jithk2000, Tony Myers, Magioladitis, Rowsdower45, PC78, BostonRed, WOSlinker, Gundam Guyver, Cheesemeister, Pammalamma, Un-registered.coward, Clockster, Lightmouse, Dravecky, Reason turns rancid, Lizziebabes90, Duvallg, ImageRemovalBot, Martarius, Kotniski, Hobocamp, Xavexgoem, Macha Panta, 718 Bot, Rogerg79, Qrusher14242, XLinkBot, Pinwiz11, Addbot, Normal View, Tide rolls, Lightbot, ShortyBGoode2020, AnomieBOT, Syco54645, Sketchmoose, Ai1238, Johnchitown, Coderjoe, Ale And Quail, RonMcAdams1, 馬格理七年, H3llBot, Rcsprinter123, Evan-Amos, Helpful Pixie Bot, Honkshoe12vv, Eegah2000, Taylor Trescott and Anonymous: 86

- **Clowns in the Sky** *Source:* https://en.wikipedia.org/wiki/Clowns_in_the_Sky?oldid=723727402 *Contributors:* Bearcat, McGeddon, Swister-Twister, Captain8track, Mattdgroves and Anonymous: 1

- **Crow T. Robot** *Source:* https://en.wikipedia.org/wiki/Crow_T._Robot?oldid=709133880 *Contributors:* Bryan Derksen, Michael Hardy, Paul A, Zoicon5, Maximus Rex, Furrykef, Jeffq, Owen, Jsonitsac, Bean shadow, Lefty, LockeShocke, Iceberg3k, Mike R, Jesster79, Kuralyov, Yos-sarian, MakeRocketGoNow, Thanos6, Robdumas, CyberSkull, Atlant, Ashley Pomeroy, Kouban, Kelly Martin, Firsfron, ItsWalky, Robert K S, Rjwilmsi, Bubba73, Sean Gray, Avalyn, VolatileChemical, Rob T Firefly, ChuckyDarko, Yamara, Hydrargyrum, Kyorosuke, BazookaJoe, KenoSarawa, H Hog, Sugar Bear, Gundam Bass, Brossow, Bluebot, D-Rock, Silent Tom, Gladrius, Unknown Dragon, Andy120290, Dread-star, Iridescence, Thistheman, Rory096, Sylocat, Trackeroc, JHunterJ, Redeagle688, Animedude360, Iridescent, Sabrewing, Tuttt, Courcelles, CmdrObot, Jlbarron, ShelfSkewed, Cydebot, Treybien, The soapboxer, Gaeamil, JAF1970, Tom servo, RobotG, Tony Myers, Trilaan, End-lessdan, CrankyScorpion, Gojirob, JamesBWatson, Kakomu, Gwern, CommonsDelinker, Drmuttonchops, Stevewills, SieBot, Allmightyduck, ImageRemovalBot, Martarius, ClueBot, David Feldmann, Tyrantulas, Arjayay, Chaosdruid, Addbot, BaerXIII, Citation bot, Inane Asylum, J929, Citation bot 1, JIK1975, Full-date unlinking bot, Golem866, Onel5969, Calizan, Jerel411, Unga Khan, ClueBot NG, IJVin, Helpful Pixie Bot, BG19bot, Dartpaw86, CetteFoisDemain, Vonran, OfTheGreen, Coratheexplorer3, Serpinium, Daftsquare, MagicatthemovieS, Tunnelofgloves and Anonymous: 101

- **Darkstar: The Interactive Movie** *Source:* https://en.wikipedia.org/wiki/Darkstar%3A_The_Interactive_Movie?oldid=724887764 *Contributors:* Piotrus, Griliopoulos, Robert K S, Bgwhite, Fram, Bdve, Quatloo, G-Flex, Mika1h, Cydebot, X201, Mudwater, Jmrowland, Comrade Graham, ImageRemovalBot, Martarius, Tassedethe, Yobot, Julle, Pixel Eater, FrescoBot, Spidey104, Mjs1991, Lotje, DASHBot, John of Reading, Jg2904, Socreamy and Anonymous: 8

- **Dr. Clayton Forrester (Mystery Science Theater 3000)** *Source:* https://en.wikipedia.org/wiki/Dr._Clayton_Forrester_(Mystery_Science_Theater_3000)?oldid=696164412 *Contributors:* Jeffq, Kuralyov, CyberSkull, Robert K S, Hailey C. Shannon, Koavf, The wub, Wack'd, Rhindle The Red, Gadget850, Nikkimaria, Th1rt3en, Sugar Bear, MrBucket, SmackBot, Hmains, Billyhart, Silent Tom, Whpq, Dream out loud, Sabrew-ing, Courcelles, Cydebot, Treybien, Thijs!bot, JAF1970, Yukichigai, Tom servo, ImABadBroth, Juliancolton, Saber girl08, Slysplace, Green-eyed girl, LeonMcNichol, ImageRemovalBot, Niceguyedc, Trivialist, Auntof6, Polly, Ariconte, Addbot, Yobot, AnomieBOT, Surv1v4l1st, JIK1975, IJVin, DoctorKubla, Mogism, Mohok, Jleevan and Anonymous: 29

- **Dr. Laurence Erhardt** *Source:* https://en.wikipedia.org/wiki/Dr._Laurence_Erhardt?oldid=679263070 *Contributors:* Jeffq, Aranel, Kjkolb, Apostrophe, Ericl234, Shanedidona, Kbdank71, Erebus555, The wub, Who, Sugar Bear, Xaosflux, Silent Tom, Dream out loud, Dreadstar, Leechcode5, David.alex.lamb, Sabrewing, Godgundam10, Cydebot, Treybien, JustAGal, Captain Crawdad, Tom servo, Slysplace, Green-eyed girl, Dravecky, Maniago, Polly, Addbot, JIK1975, IJVin and Anonymous: 23

- **Edward the Less** *Source:* https://en.wikipedia.org/wiki/Edward_the_Less?oldid=695155834 *Contributors:* Bearcat, Rhindle The Red, Closed-mouth, Mike Selinker, Cydebot, Tangurena, EagleFan, Captain Infinity, Hunter Kahn, Spidey104, Trappist the monk, John of Reading, Phnom-Pencil, BattyBot, ExpyB, Beerest355 and Anonymous: 4

- **Gypsy (Mystery Science Theater 3000)** *Source:* https://en.wikipedia.org/wiki/Gypsy_(Mystery_Science_Theater_3000)?oldid=720230170 *Contributors:* Jeffq, Astronautics~enwiki, David Gerard, Doctorcherokee, Jesster79, MakeRocketGoNow, CanisRufus, Sully, Ashley Pomeroy, Kelly Martin, Firsfron, Robert K S, Hailey C. Shannon, Chris Buckey, Koavf, Asciic, RevRaven, Rhindle The Red, Thatdog, Syrthiss, Iwant2baurora, Betacommand, Silent Tom, Andy120290, Larry Hastings, Leechcode5, Jwadeg, Sabrewing, Courcelles, Drinibot, Cydebot, Treybien, Thijs!bot, JAF1970, Sochwa, Tom servo, RobotG, MegX, Maias, ZPM, Mnpeter, Rwe1138, RJASE1, Slysplace, Tyrantulas, Addbot, Yobot, J929, FrescoBot, JIK1975, IJVin, BeanoMaster, MaxamillionSmart, Nin26plus and Anonymous: 24

- **Joel Robinson** *Source:* https://en.wikipedia.org/wiki/Joel_Robinson?oldid=720445161 *Contributors:* Dascott, Mulad, Jeffq, Iceberg3k, Mike R, Copperboom, Jesster79, Kuralyov, Wasabe3543, MakeRocketGoNow, Rmsharpe, Goatbear, Ashley Pomeroy, Zee~enwiki, Duke33, Sabo,

Trogga, Fred J, John Anderson, VolatileChemical, Wack'd, Daverocks, ChuckyDarko, Kyorosuke, Cleared as filed, TimeDoctor, Basilsblog, Sugar Bear, Eptin, Fireballil, Skyrocket, OrphanBot, Silent Tom, Andy120290, Pissant, Gobonobo, JHunterJ, Imagine Wizard, Sabrewing, Dk1965, Cydebot, Treybien, JAYMEDINC, PKT, RobotG, Huphelmeyer, Appraiser, Trebligoniqua, Clerks, Slysplace, Samisweirdout, Krucz36, LeonMcNichol, The Thing That Should Not Be, Maniago, Polly, Wolfer68, WikHead, Addbot, TySoltaur, Tassedethe, Yobot, Materialscientist, FrescoBot, JIK1975, Dsavage87, GoingBatty, Solarra, ClueBot NG, IJVin, Godzilla and mst3k, MaxamillionSmart and Anonymous: 40

- **List of Mystery Science Theater 3000 characters** *Source:* https://en.wikipedia.org/wiki/List_of_Mystery_Science_Theater_3000_characters?oldid=712817070 *Contributors:* Bearcat, Rich Farmbrough, Redfarmer, Woohookitty, Wack'd, SmackBot, Courcelles, Jedzz, Alaibot, PKT, Nick Number, WarddrBOT, CultureDrone, ImageRemovalBot, Addbot, Friginator, Tassedethe, NeoBatfreak, AnomieBOT, Ulric1313, JIK1975, WildBot, Smartie2thaMaxXx, BG19bot, Khazar2, Ranze, MaxamillionSmart and Anonymous: 4

- **List of Mystery Science Theater 3000 episodes** *Source:* https://en.wikipedia.org/wiki/List_of_Mystery_Science_Theater_3000_episodes?oldid=724978069 *Contributors:* Jeandré du Toit, Charles Matthews, Eliot~enwiki, Jeffq, Bearcat, Nufy8, Dina, Timvasquez, Dbenbenn, Gtrmp, Lefty, AlistairMcMillan, Iceberg3k, Bobblewik, Tagishsimon, Junkyardprince, Calm, Mike R, Jesster79, Sam Hocevar, Joyous!, Tromatic, MakeRocketGoNow, DMG413, Acsenray, The stuart, D6, Rich Farmbrough, Vague Rant, Antaeus Feldspar, Blade Hirato~enwiki, Bender235, MattTM, Crooow, Cmdrjameson, Benfergy, Thanos6, Maxdillon, BillCook, Jason One, Proteus71, Ricky81682, Orelstrigo, Neilmckillop, Crablogger, Gunter, Ringbang, Kazvorpal, Kamezuki, Gmaxwell, Woohookitty, Bjones, Bellhalla, DoctorWho42, Fred J, Kanenas, Marudubshinki, Volatile, Mandarax, Graham87, BD2412, Roger McCoy, Imperialles, Tim!, Koavf, Lockley, CR85747, Wowbobwow12, Bgwhite, Wack'd, Wasted Time R, EamonnPKeane, RussBot, DarkfireTaimatsu, CanadianCaesar, Shawn81, IanManka, Gaius Cornelius, Jenblower, NateDan, Brooza, Rhindle The Red, Burrito, Wiki alf, Thalter, Cholmes75, InvaderJim42, ScottyWZ, Mlouns, Mole Man, Ospalh, Redward1, EmiOfBrie, Basilsblog, Amccaf1, Pegship, American2, Kernunrex, Closedmouth, Sugar Bear, Mightyhog, Fireballil, That Guy, From That Show!, SmackBot, Nfitz, Simon Beavis, Bobzchemist, Brossow, Gjs238, HalfShadow, Vanhagar3000, Hmains, Bluebot, Fuzzform, Alden Bates, Toughpigs, Sullenspice, Lenin and McCarthy, Silent Tom, Andy120290, Jackohare, Larry Hastings, Ithizar, Ohconfucius, Majorclanger, UKER, Bawtyshouse, Cratylus3, Peyre, Skbarton13, Marshall Stax, Nehrams2020, Donmac, Sabrewing, Prisonerdrw, George100, CmdrObot, Jedzz, ShelfSkewed, Ballista, Cassmus, Ken Gallager, J-boogie, Fred8615, Cydebot, Hebrides, DrJohnnyDiablo, After Midnight, Dtgriscom, DanDud88, JustAGal, SickBoy, Kbthompson, Dalgspleh, Pixelface, Tony Myers, Hypnometal, MegX, Andylindsay, Bakilas, Wrightaway, Hoverfish, Kaijucole, Danleary25, Mermaid from the Baltic Sea, SuperMarioMan, Evil-yuusha, Richiekim, VitaleBaby, Ownage2214, P4k, JohnnyRush10, Largoplazo, The Legend of Julie Egbert, MArcane, Black Kite, Cheesemeister, Slysplace, Arkyopterix, Bladez636, RadiantRay, Sb2007, Stigmar the destroyer, SieBot, DX927, RARoth, Dravecky, Reason turns rancid, Zyrkh, JustinT1977, Pinkadelica, ImageRemovalBot, ClueBot, Tom H12, EoGuy, FieldMarine, Mild Bill Hiccup, Trivialist, Maniago, Mpmcarthur78, Polly, Zombie Hunter Smurf, TRTX, Good Olfactory, Addbot, Friginator, Tassedethe, AnomieBOT, Utternerd, LilHelpa, COGInk, Betty Logan, Starman15317, Eugene-elgato, Perks71, Snort Barfly, FrescoBot, Surv1v4l1st, Fortdj33, Mayo15, Ravendrop, Wildhoney66, Redrose64, LizzieBabes419, Johnchitown, Notundercovercop327, ScottMHoward, Theburn77, Chmg1, Tbhotch, Ben Bohn, Torchwoody, John of Reading, GoingBatty, IncD, Metalclerk, Westholmes2001, LoremIpsumDolorSitAsmet, Zaher.Kadour, Takeyo, Ben Bohn 89, Smartie2thaMaxXx, Jalexander-WMF, Unga Khan, Targaryen, ClueBot NG, IJVin, Easy4me, Msampo, Vincelord, DanielWaste, Tfmisc, Scottwindcrest, ElliotRosewater, TheMikeBlackSpecial, SWStiletto, Mogism, The Jester In Orange, Fireflyfanboy, Frosty, Theskinnytypist, Taylor Trescott, Stephanie.molnar, Eman235, JasSpy, MaxamillionSmart, Snoopy012, Kekkomereq1 and Anonymous: 282

- **List of RiffTrax** *Source:* https://en.wikipedia.org/wiki/List_of_RiffTrax?oldid=724882190 *Contributors:* Jeffq, Mike R, Kuralyov, Discospinster, Rich Farmbrough, Journ, LtNOWIS, BathTub, Woohookitty, BD2412, Koavf, Alaney2k, EamonnPKeane, Sillstaw, Rob T Firefly, RadioFan2 (usurped), Number 57, Sugar Bear, Doc Strange, Hmains, NES Boy, Tv's emory, Kyojikasshu, Jedzz, A876, HauntingYourKids, PamD, DanTD, Nicholas0, Fayenatic london, Blakestern, TAnthony, Bakilas, KConWiki, General Jazza, Richiekim, Bovineboy2008, Paradoctor, Lizziebabes90, K1posterchild, Martarius, EoGuy, Niceguyedc, Sun Creator, Yonskii, Zombie Hunter Smurf, Jafeluv, C1c9k72, Blethering Scot, Tassedethe, Yobot, Gongshow, Sonam8311, Emanresu27, Kikisfishy, Starman15317, FrescoBot, JIK1975, DrilBot, Spidey104, Kelseypedia, Dsavage87, John of Reading, Stryn, Fandraltastic, Moswento, Dagko, Deeez Nuuuts, Davrane, PlaidFlannelShirt, Jscottb805, ChrisGualtieri, Mike.Dibble, MATERIALflows, Mogism, PeaceShield5, JRicker,PhD, Svengali Rising, TheCartographer72, MoeIsMe, Beerest355, Liz, Taylor Trescott, Gordiac, SillstawT, TheMagikCow, Darkenedsky, Debbiesw, Toplar Scarre and Anonymous: 162

- **Mike Nelson (character)** *Source:* https://en.wikipedia.org/wiki/Mike_Nelson_(character)?oldid=724072184 *Contributors:* Furrykef, Kaal, Jeffq, David Gerard, Iceberg3k, Mike R, Jesster79, Kuralyov, Wasabe3543, Ashley Pomeroy, Lectonar, Woohookitty, Hailey C. Shannon, VolatileChemical, Wack'd, Wavelength, Gyre, Aussie Evil, Kyorosuke, PrimeCupEevee, SmackBot, Felicity4711, Brossow, Bluebot, Fuzzform, Skyrocket, Oatmeal batman, Lenin and McCarthy, TechPurism, Bando26, ShaleZero, Sabrewing, Courcelles, Cydebot, JAF1970, JustAGal, Tom servo, RobotG, Ajb333, Lightwing1988, Rwe1138, Trumpet marietta 45750, Skier Dude, Slysplace, LeonMcNichol, ImageRemovalBot, Ideabook, Maniago, Addbot, Erik9bot, Surv1v4l1st, JIK1975, Haon 2.0, IJVin, BG19bot, PhnomPencil, Superhollyhox, Khazar2, Beerest355 and Anonymous: 38

- **MSTing** *Source:* https://en.wikipedia.org/wiki/MSTing?oldid=715120238 *Contributors:* Karen Johnson, Furrykef, Jeffq, Lefty, Jesster79, SAMAS, Kuralyov, Rich Farmbrough, Clawed, Antaeus Feldspar, Thu, Crowbar~enwiki, Cmdrjameson, Ununnilium, Zeborah, Chardish, Veemonkamiya, Robert K S, Hbdragon88, Liface, Marudubshinki, BD2412, Teflon Don, Koavf, Kerowyn, Celestianpower, Gurch, Hibana, Rebochan, Rob T Firefly, Spike Wilbury, Robertvan1, AlexMW, Retired username, Rouge2, Rayc, Selmo, Gundam Bass, SmackBot, David Kernow, Malkinann, The Famous Movie Director, Toughpigs, Andy120290, Fuhghettaboutit, EVula, Runa27, Masem, Sabrewing, CmdrObot, Casper2k3, Ollie Garkey, Meinterrupted, Luigifan, Haha169, Deflective, ChazBeckett, Wyatt Payne, Neilrickaby, Slysplace, Moonriddengirl, Lizziebabes90, YSSYguy, EoGuy, Wysprgr2005, Bill l rr, TBustah, PrincessLeiaOrgana, Yonskii, WPjcm, Gongshow, AnomieBOT, LilHelpa, Full-date unlinking bot, Lightlowemon, Updatehelper, Mikemacdee, H3llBot, BornonJune8, ClueBot NG, Grimm67, PhnomPencil, Cyberbot II, Amortias, SecretPedia, Zoogz and Anonymous: 60

- **Mystery Science Theater 3000** *Source:* https://en.wikipedia.org/wiki/Mystery_Science_Theater_3000?oldid=724841119 *Contributors:* Bryan Derksen, Dascott, Modemac, Paul A, Goatasaur, Tregoweth, Potatoscone, Lee M, Ghewgill, Wfeidt, Arteitle, Akira742, Mulad, Mw66, WhisperToMe, Zoicon5, Furrykef, Saltine, Martin Edelius, JonathanDP81, Jeffq, Sjorford, Dale Arnett, Xuanwu, RedWolf, Chocolateboy, Modulatum, Chris Roy, Postdlf, Tualha, Lesonyrra, Jsonitsac, Mr w~enwiki, Bean shadow, Timvasquez, DocWatson42, Inter, Wwoods, Everyking, Lefty, NeoJustin, Varlaam, BigHaz, Mboverload, Iceberg3k, Bobblewik, Mike R, MisfitToys, Copperboom, Jesster79, Kuralyov, Rlcantwell, WOT, Wasabe3543, Zondor, Jrp, Discospinster, Rich Farmbrough, Huffers, Ahkond, Horkana, Antaeus Feldspar, Andrew Maiman, Lachatdelarue,

Bender235, Jnestorius, Crooow, Drted, Spearhead, TMC1982, Causa sui, Deathawk, Bustter, Walkiped, BalooUrsidae, Thanos6, Maxdillon, Forteanajones, Jason One, Corax, Storm Rider, Proteus71, Diego Moya, Nurban, Andrewpmk, Ricky81682, Bblackmoor, Ashley Pomeroy, Sade, Redfarmer, Roadrunner3000, InShanee, Circuitloss, Ronark, TheRealFennShysa, Zee~enwiki, Sketchee, Drat, Outlanderssc, Kouban, Kazvorpal, Kitch, Duke33, Lkinkade, Chardish, Bacteria, Firsfron, Woohookitty, Lydia Pryon, DoctorWho42, Robert K S, Fred J, Hailey C. Shannon, Hbdragon88, Chris Buckey, Andreas -horn- Hornig, Zzyzx11, Leemeng, Hooperx, Pfalstad, Dbutler1986, Lego872, Barvobot, Rjwilmsi, Koavf, Tuntis, Bubba73, Tarc, Horseytown, Chromium.switch, Yamamoto Ichiro, JohnDBuell, A scientist, Nihiltres, Azezel, Fragglet, RexNL, Gurch, RevRaven, Nivaca, Jonny2x4, Wowbobwow12, Jfiling, Thegreatmonkey, Sharkface217, Mhking, Kellywatchthestars, Thorshammer4283, VolatileChemical, Bgwhite, Rebochan, Jason.cinema, Wack'd, EamonnPKeane, YurikBot, Quentin X, Rob T Firefly, Ehdee, Aussie Evil, Kymacpherson, Cliffb, WAvegetarian, SluggoOne, Tenebrae, NateDan, Big Brother 1984, Rhindle The Red, ONEder Boy, Daikiki, Nutiketaiel, AlexMW, Dureo, Irishguy, Ragesoss, Marvin01, Doctorindy, InvaderJim42, Bobak, Ma3nocum, EmiOfBrie, Furiouszebra, Black Falcon, Basilsblog, Amccaf1, Dan Harkless, Capt Jim, John Pannozzi, Kernunrex, Conan-san, Nikkimaria, Th1rt3en, YoungAmerican, MStraw, Mad283, Bcarlson33, Wainstead, Thephotoplayer, Sugar Bear, Aeolis7, John Broughton, Eptin, ParticularlyEvil, Fireballil, Dposse, Burnwelk, SmackBot, Franny Wentzel, McGeddon, WikiuserNI, Brossow, Hmains, Wieners, ERcheck, Katanin, Oneismany, Dan Hoey, AshuraH, BullWikiWinkle, Asclepius, Thumperward, Happylobster, Alumni, Gareth, Vanyel, Rolypolyman, Colonies Chris, Sullenspice, Huwmanbeing, Can't sleep, clown will eat me, MisterHand, Sixofone88, OrphanBot, Jennica, IQpierce, Maetch, Matthew, Azumanga1, Konczewski, Fightingirish, TKD, Gladrius, Andy120290, Paradox CT, Cornprone, Jmlk17, KnowBuddy, DerHerrMigo, Downtown dan seattle, Chaos386, EVula, Hoof Hearted, Chargh, Grejlen, Larry Hastings, Ithizar, Sonic Hog, Salamurai, Ceoil, Ohconfucius, Maddogmike, ArglebargleIV, Sylocat, Attys, TomRK1089, T-dot, SweetHeart666, Adavidw, Dynayellow, Bando26, InsaneZeroG, Shadowlynk, Leechcode5, RowanInBlack, ExtraordinaryMan, Polyhymnia, Majorclanger, JerryLewisOverdrive, Ckatz, A. Parrot, Intersting, Loadmaster, Shamrox, Gordonov, Beetstra, Boomshadow, Fearghas K, Redeagle688, Peyre, Masem, MikeWazowski, Marshall Stax, Levineps, Jwadeg, Nehrams2020, Iridescent, GrandpaPap, Donmac, Danger bird, Sabrewing, GrizzlyFlats, Threepwood7, Filliam H Muffman, Kyojikasshu, Mrquizzical, Denzilq, Woodshed, George100, Davidbspalding, ChrisCork, PurpleRain, Idols of Mud, Jlbarron, Wafulz, Jedzz, Picaroon, Jkazoo, Kylu, ThisIslandEarth, ShelfSkewed, Dairhenien, MarsRover, K00bine, Casper2k3, The Enslaver, Fred8615, Moofoo, Cydebot, Erasmussen, Otto4711, ST47, Weezerzero, DrJohnnyDiablo, Pufnstuf, Plasticbadge, M.S.K., RedWolfX, Jeff.Mortimer, Some Person, Numen, JAF1970, Behrens64, Wikid77, Umsner, Stevekl, Yukichigai, Eco84, Nutmilk, ThatGuamGuy, John254, Wxyzone, Zachary, Tom servo, Rjmars97, AntiVandalBot, Fireplace, MoogleDan, Luna Santin, Droidguy1119, CobraWiki, SeraphisCain, RobJ1981, Recharge138, ImpossibleEcho, UncolaMan, NapoliRoma, Tony Myers, Chem242, Sonicsuns, Natebell, Crazyboy899, Lightwing1988, TAnthony, MegX, Andylindsay, Delius1967, Mr. Erik, Steveprutz, Magioladitis, PacificBoy, Dp76764, LafinJack, Prestonmcconkie, Satch234, EagleFan, MetsBot, Hoverfish, Kiminatheguardian, Kaijucole, Dinot403, Mermaid from the Baltic Sea, Akbeancounter, John Doe or Jane Doe, Chewieshmoo, SuperMarioMan, Griglager, Evil-yuusha, Mausy5043, Tommy11111, Benscripps, Clerks, Katharineamy, Mdumas43073, AntiSpamBot, Vanished user g454XxNpUVWvxzlr, NewEnglandYankee, DiscordantNote, JohnnyRush10, Largoplazo, Lossleaders, Tkgd2007, Xiahou, Mckaycr, MArcane, Lights, 28bytes, Sjones23, Charlycrash, DancingMan, Klaatus-Robotman, Sam444444, Barneca, JisforJoe, Neilrickaby, Lots42, Cheesemeister, NPrice, JayJayTheSpitfire, Sam927, Imasleepviking, Slysplace, Amis2007, ^demonBot2, Jamesfett, Agentgreen004, Dan Cziraky, Wiae, Onore Baka Sama, MrTorso, LanceBarber, Impasse, Jhawkinson, Samisweirdout, Genericface, Grief23, SparklingJoe, Thatother1dude, Km9000, SieBot, BlankHole, Unregistered.coward, Jason Patton, Macgyver89, Captain Yankee, Editore99, Eóin, Tritium h3, Rosspz, Joeartguy, Yerpo, Turtlescrubber, Wiki-maann, Fuddle, HighInBC, Altzinn, Pinkadelica, Revelian, Aaron045, ImageRemovalBot, SkyGuySSS, Wtvg1, Deanlaw, Theseven7, Pwitham, Brian Siano, Mild Bill Hiccup, Niceguyedc, Tubafishy, Bjeggert82, Luckibrian, Midwestmystie, Three-quarter-ten, Nin10doh, Doc cromwell, 12 Noon, BondoFox, Arjayay, Petevo, Polly, Mlaffs, Salon Essahj, Yonskii, Bjdehut, DumZiBoT, InternetMeme, Thinkandsuggest, Zombie Hunter Smurf, XLinkBot, King of kod, Kytti khat, Eblackadder3, Hunter Kahn, Kbdankbot, Addbot, BarberFett, Landon1980, Friginator, Sabianinnc, Leszek Jańczuk, Musdan77, JPMST, TySoltaur, Tassedethe, 84user, Killy mcgee, Luckas-bot, Yobot, Themfromspace, MechaValis, Kjell Knudde, Gongshow, Athelstanent, AnomieBOT, Jclifford83, Utopian Laser Beam, Marsdemartini, Kingpin13, Ulric1313, Utternerd, Ckruschke, Citation bot, Person13, LilHelpa, Sketchmoose, Millahnna, Ariel19, Ansonite, Starman15317, Riotrocket8676, Pixel Eater, Ballyhoofilms1, GT5162, J929, FrescoBot, Surv1v4l1st, Fortdj33, Nkratter, Ravendrop, Anaphysik, Citation bot 1, JIK1975, Jonesey95, LizzieBabes419, Tom.Reding, Patrick McDougle, Pilot 2023, Madnana42, Lotje, Golem866, Jmaxwell2400, Woodlot, RjwilmsiBot, Mpompu, Burmiester, WildBot, EmausBot, John of Reading, Lunaibis, Haon 2.0, Dewritech, GoingBatty, Bt8257, Italia2006, Jg2904, ZéroBot, Daonguyen95, NathanielTheBold, Lordmas, H3llBot, SporkBot, Wenttomowameadow, BNSF1995, SpencerCollins, Rcsprinter123, Jay-Sebastos, Peckoduck, BookDen, Edfan1, Smartie2thaMaxXx, BornonJune8, Goldiegopher, Terraflorin, Asuccesfulguy, Burtnoternie, ClueBot NG, Jackstikishack, Joefromrandb, Jamo58, Vincelord, Harley Hudson, Zenasdude, Myuphrid, BG19bot, Neptune's Trident, Kixon66, Mikel travis, Geraldo Perez, Snow Rise, Arcolye, Laodah, Cyberbot II, Khazar2, Hindumuninc, EditorE, EagerToddler39, FiverFan65, Mogism, The Jester In Orange, Fireflyfanboy, Partyclams, Yoshiman6464, TwoTwoHello, Glennedward, Jodosma, Lemaroto, Serpinium, Mr. Lama, Beerest355, Joevietman, Billiejean89, Taylor Trescott, MagicatthemovieS, Sibtye, Bieler0317, ColRad85, Monkbot, Vieque, WikiOriginal-9, Flargg, Zorono Ornitorrico, Justbecause5, Nohomersryan, Oiyarbepsy, JShanley98, FriarTuck1981, Beardem36, MaxamillionSmart, Paredokz, Manowich, Popnerd, Snoopy012, ElizaLepine, ChimpyJoe, Archiejames, Ripley 101, Kekkomereq1, Colonel Wilhelm Klink, GreenC bot and Anonymous: 793

- **Mystery Science Theater 3000 (Flash series)** *Source:* https://en.wikipedia.org/wiki/Mystery_Science_Theater_3000_(Flash_series)?oldid= 675508456 *Contributors:* Mike R, Thanos6, Rjwilmsi, Koavf, Savethemooses, CR85747, Wack'd, Rob T Firefly, SmackBot, Andy120290, Shamrox, Boomshadow, GrizzlyFlats, CmdrObot, Thijs!bot, Oxguy3, AntiSpamBot, Lizziebabes90, ImageRemovalBot, Trivialist, XLinkBot, Addbot, TutterMouse, FrescoBot, Surv1v4l1st, JIK1975, LizzieBabes419, BookDen, StuPeg, ~riley and Anonymous: 14

- **Mystery Science Theater 3000 home video releases** *Source:* https://en.wikipedia.org/wiki/Mystery_Science_Theater_3000_home_video_ releases?oldid=724638298 *Contributors:* Jeffq, Patsw, Kouban, Kbdank71, CR85747, Wack'd, EamonnPKeane, Musicpvm, DarkfireTaimatsu, NateDan, Basilsblog, Sugar Bear, DT29, SmackBot, Chris the speller, MisterHand, Harryboyles, J 1982, Mark Lungo, Texas Dervish, Marshall Stax, Jedzz, ShelfSkewed, Otto4711, DrJohnnyDiablo, After Midnight, JustAGal, Waacstats, Rowsdower45, Hbent, OutsiderT, Signalhead, Slysplace, Finngall, Samisweirdout, JustinT1977, ImageRemovalBot, ClueBot, EoGuy, Mortense, BarberFett, Friginator, Krthomas, Tassedethe, Yobot, Ulric1313, LilHelpa, Ariel19, Hi878, Starman15317, FrescoBot, Jedicrippler, Johnchitown, Notundercovercop327, Jlouise311, Retropolis1, Coasterlover1994, Harley Hudson, TVWolf, Mark Arsten, CAWylie, Dexbot, Fireflyfanboy, Beerest355, T.Y. Faltermeyer, Naicnayrb, Zyrkhh, MaxamillionSmart, Nphieb, Academy90210, Jonasrosland, Kekkomereq1 and Anonymous: 77

- **Mystery Science Theater 3000: The Movie** *Source:* https://en.wikipedia.org/wiki/Mystery_Science_Theater_3000%3A_The_Movie?oldid= 724961942 *Contributors:* Frecklefoot, Paul A, Tregoweth, Jeffq, David Gerard, Gamaliel, Xinoph, Iceberg3k, Mike R, Jesster79, Sam Hocevar, MakeRocketGoNow, Bender235, Jason One, Crablogger, Erik, Woohookitty, BD2412, Rjwilmsi, Tdowling, MarnetteD, Yamamoto Ichiro, CR85747, RevRaven, YurikBot, SpikeJones, Sceptre, Rob T Firefly, Tenebrae, Pegship, Capt Jim, Conan-san, JQF, Sugar Bear, Van Hagar, Timothy da Thy~enwiki, Fireballil, SmackBot, Brossow, BiT, Lenin and McCarthy, Maetch, Matthew, TKD, Andy120290, Ser Amantio di Nicolao, Sylocat, Leechcode5, Boomshadow, Whomp, Locutus, Zepheus, Masem, Acaudel, Sabrewing, Esn, Jedzz, Cydebot, Lonenut2000, Mallanox, JAF1970, Umsner, QuasyBoy, JustAGal, Gunnafan, Witteafval, Sreejithk2000, Andrzejbanas, Tagenar, Bencherlite, Magioladitis, DXRAW, Wrightaway, MartinBot, Benscripps, DOHC Holiday, Slysplace, Polbot, JohnnyMrNinja, SkyGuySSS, David Feldmann, Trivialist, Maniago, Cliff1911, I am the Paulrus, Kbdankbot, Addbot, Friginator, Yobot, CarpetCrawler, AnomieBOT, Ulric1313, TVsEgon, JDLewis007, JIK1975, Pawsrent, GoingBatty, JDDJS, DeWaine, Ὁ οἶστρος, SporkBot, Terraflorin, 19jduryea, BG19bot, PhnomPencil, Gabriel Yuji, TF-Syndicate, Billiejean89, MagicatthemovieS, Zorono Ornitorrico, JShanley98, SpaceGoofsGeekerBoy and Anonymous: 84

- **Observer (Mystery Science Theater 3000)** *Source:* https://en.wikipedia.org/wiki/Observer_(Mystery_Science_Theater_3000)?oldid= 666133809 *Contributors:* Fibonacci, Jeffq, Iceberg3k, Jesster79, Kuralyov, MakeRocketGoNow, Grstain, Jason One, SemperBlotto, P Ingerson, Ceyockey, Woohookitty, Robert K S, Koavf, Jonny2x4, Wack'd, EmiOfBrie, Sugar Bear, Brossow, Silent Tom, Pookster11, Letoofdune, Sabrewing, George100, Cydebot, JAF1970, Multiverse, Tom servo, NapoliRoma, Slysplace, ThorZero, Green-eyed girl, Maniago, Schreiber-Bike, Polly, Fortdj33, JIK1975, John of Reading, Haon 2.0, IJVin, Braincricket, ChrisGualtieri, DoctorKubla and Anonymous: 19

- **Pearl Forrester** *Source:* https://en.wikipedia.org/wiki/Pearl_Forrester?oldid=719191715 *Contributors:* Jeffq, Robbot, Jesster79, Kuralyov, MakeRocketGoNow, Jason One, Kitch, Robert K S, Rjwilmsi, Gsp, Wack'd, Aussie Evil, Dankstick, Amccaf1, Smurrayinchester, Sugar Bear, SmackBot, Brossow, Silent Tom, Disavian, Doczilla, Sabrewing, ShelfSkewed, Cydebot, JAF1970, Tom servo, MegX, Clerks, Charlycrash, Green-eyed girl, Polly, Addbot, JIK1975, Celtic mama, SporkBot, IJVin, BattyBot, DoctorKubla, Me, Myself, and I are Here and Anonymous: 25

- **Professor Bobo** *Source:* https://en.wikipedia.org/wiki/Professor_Bobo?oldid=666134636 *Contributors:* Jeffq, Iceberg3k, Jesster79, Ganymead, Kuralyov, MakeRocketGoNow, Deelkar, Che fox, Thanos6, Brother Dave Thompson, Jason One, Woohookitty, Trogga, BD2412, Wack'd, DarkfireTaimatsu, GusF, Gaius Cornelius, Nikkimaria, Sugar Bear, Brossow, Sadads, Silent Tom, Redeagle688, Sabrewing, Cydebot, Treybien, JAF1970, Darkstaruav, Tom servo, MegX, Clerks, Slysplace, BrianAdler, Green-eyed girl, Goustien, KathrynLybarger, Maniago, Polly, Krumovies26, Addbot, The Robot 2000, Fortdj33, JIK1975, IJVin, DoctorKubla and Anonymous: 15

- **RiffTrax** *Source:* https://en.wikipedia.org/wiki/RiffTrax?oldid=720942910 *Contributors:* Conti, Jfruh, DocWatson42, Mike R, Kuralyov, Rich Farmbrough, TMC1982, Drat, Krellion, Kouban, Woohookitty, Fred J, Iisryan, BD2412, Kbdank71, Roger McCoy, Nightscream, Koavf, CR85747, DarKrow, Wowbobwow12, Gplefka, Hawaiian717, Rob T Firefly, InvaderJim42, William Graham, Amccaf1, MarsJenkar, Sugar Bear, SmackBot, Jkp1187, Lamerc, Tjwagner, Rolypolyman, Toughpigs, NYKevin, Egsan Bacon, Silent Tom, Dmoon1, Andy120290, NES Boy, Ithizar, Salamurai, Byelf2007, Jzummak, Bando26, Beetstra, Redeagle688, Doczilla, Masem, Sabrewing, Casper2k3, UberMan5000, Thijs!bot, Andyjsmith, DanTD, Nick Number, Mdriver1981, Jhsounds, Graveenib, Steveprutz, Kgagne, Badfan, Rowsdower45, Chrombot, Antmusic, Islandersa, Richiekim, Tommy11111, Antepenultimate, Wikimandia, MorseMoose, McNoddy~enwiki, Scottrandall, Lexein, Charlycrash, Neilrickaby, Togedude, TheValentineBros, Jhawkinson, Mcgonigle, Unregistered.coward, BPlague, Ruiner2001, ImageRemovalBot, ClueBot, Lazlo25, Niceguyedc, Arjayay, JustinStolle, JasonAQuest, Chetpooner, InternetMeme, XLinkBot, Addbot, BarberFett, Normal View, Tassedethe, Luckas-bot, Yobot, Sonam8311, AnomieBOT, Jim1138, Meattrademark, Venice85, Starman15317, Selk1138, JIK1975, Trappist the monk, RjwilmsiBot, RonMcAdams1, John of Reading, Aburne45, GoingBatty, Grigorisio, Bryce Carmony, H3llBot, SteveScripps, ReiCow, BG19bot, YautjaVeteranWolf, BattyBot, Rockinlobster, Beerest355, NCFan12312, Sa121sBu, SillstawT, JShanley98, JLHockeyKnight, Nphieb, Snoopy012, Ragaboo3, Awcdpazz89, Toplar Scarre and Anonymous: 138

- **Satellite of Love (Mystery Science Theater 3000)** *Source:* https://en.wikipedia.org/wiki/Satellite_of_Love_(Mystery_Science_Theater_3000) ?oldid=703318782 *Contributors:* Bryan Derksen, Mulad, Eugene van der Pijll, Jeffq, Iceberg3k, Mike R, Jesster79, Wasabe3543, Thanos6, Jason One, Mithent, Kelly Martin, Woohookitty, Vash The Stampede, Rjwilmsi, Koavf, Horseytown, Gurch, Wack'd, Pigman, Welsh, Tokachu, Mole Man, EmiOfBrie, Amccaf1, SmackBot, Felicity4711, GoodDay, Tschwenn, Neo139, Silent Tom, Andy120290, KnowBuddy, Leechcode5, Peyre, −5-, Sabrewing, Lenoxus, Courcelles, Jlbarron, Djcastel, Lonenut2000, ST47, Rspeed, Tom servo, JamesBWatson, Ekotkie, Jeepday, Zakuragi, Slysplace, Lightmouse, 47of74, Maniago, Addbot, Najhoant, Unimath, J929, JIK1975, Animalparty, Vercatosso, Dewritech, Bt8257, Jerel411, Helpful Pixie Bot, Harley Hudson, ChrisGualtieri, Khazar2, DavidLeighEllis, MagicatthemovieS and Anonymous: 34

- **The Film Crew** *Source:* https://en.wikipedia.org/wiki/The_Film_Crew?oldid=719007719 *Contributors:* Tregoweth, Jeffq, Jason One, Kouban, Woohookitty, Koavf, The wub, Rob T Firefly, RussBot, Robertvan1, Closedmouth, Sugar Bear, SmackBot, Nsayer, Brossow, Chris the speller, Salamurai, Cydebot, Bddmagic, JamesAM, Thijs!bot, Kgagne, P64, Yoni, Akbeancounter, Bovineboy2008, JimRH123, Lizziebabes90, TubularWorld, Duvallg, Yonskii, XLinkBot, Addbot, Normal View, Meattrademark, BG19bot, ChrisGualtieri, TheHummingbird02 and Anonymous: 26

- **Timmy Big Hands** *Source:* https://en.wikipedia.org/wiki/Timmy_Big_Hands?oldid=695155891 *Contributors:* Tregoweth, Jeffq, DragonflySixtyseven, Deathawk, Ashley Pomeroy, Robert K S, Koavf, Seraphimblade, Wack'd, Gaius Cornelius, Irishguy, Anetode, DeadEyeArrow, Ms2ger, Brossow, Kuru, Thedp, Crowish, Alaibot, TangentCube, AntiVandalBot, SeraphisCain, MartinBot, TomasBat, Qwertyyqwe, Slysplace, Team underage, Lextil259, Cooluser607, Oda Mari, BobbyBobBobBobBobBOB, Trivialist, DumZiBoT, Lightbot, HCShannon, K6ka, ClueBot NG and Anonymous: 15

- **Tom Servo** *Source:* https://en.wikipedia.org/wiki/Tom_Servo?oldid=704095622 *Contributors:* Bryan Derksen, Paul A, Furrykef, Jeffq, Bearcat, Meelar, Bean shadow, Lefty, Mboverload, Iceberg3k, Mike R, Jesster79, Kuralyov, MakeRocketGoNow, Ahkond, Thanos6, Robdumas, Ashley Pomeroy, Crystalllized, Kouban, Duke33, SDC, BD2412, Rjwilmsi, Koavf, Bubba73, Supermorff, RevRaven, Wack'd, Rob T Firefly, ChuckyDarko, Shaddack, Kyorosuke, NateDan, Nikkimaria, Th1rt3en, Sugar Bear, Aeolis7, Fireballil, SmackBot, Brossow, Betacommand, Jnelson09, Droll, Toughpigs, George Ho, Silent Tom, Konczewski, Andy120290, PsychoJosh, Ck lostsword, Leechcode5, Big Smooth, Animedude360, Sabrewing, George100, J Milburn, Cydebot, Treybien, Gaeamil, JAF1970, Tom servo, RobotG, ReverendG, Nevlik, Tony Myers, Steveprutz, SLWalsh, Ihnatko, Trevor Burnham, CommonsDelinker, Benscripps, Captain Infinity, Kinkyfish, RJASE1, Hirolovesswords, Eggman1115, Mmacnair, Mplsray, Goatonastik, Slysplace, Holiday56, Fratrep, ImageRemovalBot, The Thing That Should Not Be, Tigerboy1966, Tyrantulas, Beadbop, Maniago, Zombie Hunter Smurf, Addbot, Gongshow, Lightningbarer, J929, FrescoBot, Surv1v4l1st, Skyerise, Dewritech, Faolin42, Starcheerspeaksnewslostwars, Ὁ οἶστρος, Weegee34, IJVin, Jamo58, SteenthIWbot, Triplecrown120, Philip Sisk, MaxamillionSmart, PrimeTheMexicanAthiest and Anonymous: 99

- **TV's Frank** *Source:* https://en.wikipedia.org/wiki/TV'{}s_Frank?oldid=693145676 *Contributors:* Jeffq, Lefty, Iceberg3k, Jesster79, Kuralyov, MakeRocketGoNow, Aranel, Thanos6, Robdumas, Jason One, Ceyockey, Hailey C. Shannon, A Train, BD2412, RevRaven, Ehdee, Amccaf1, Sugar Bear, SmackBot, Hmains, Silent Tom, EVula, G33K, Loadmaster, Shamrox, Redeagle688, Sabrewing, Cydebot, Treybien, PKT, Yu-kichigai, X96lee15, Tom servo, Skier Dude, Slysplace, Handarazuur, Km9000, Green-eyed girl, LeonMcNichol, ImageRemovalBot, Sinandls, Maniago, Polly, Yobot, EmausBot, IJVin, DoctorKubla and Anonymous: 37

- **WUCW** *Source:* https://en.wikipedia.org/wiki/WUCW?oldid=716621299 *Contributors:* Mulad, Smith03, DJ Clayworth, Jeffq, Skybunny, Davodd, HangingCurve, LarryGilbert, Xinoph, Iceberg3k, Jesster79, D6, Discospinster, Rich Farmbrough, Mwmnp, Evice, Rockhopper10r, Stephen Bain, Leonard23, Ianblair23, Boothy443, Hailey C. Shannon, BD2412, Vegaswikian, CR85747, RobyWayne, RevRaven, Mrschimpf, Kjammer, Wack'd, Wavelength, CFIF, ScrippsONEDetroit, Rhindle The Red, Tvtonightokc, EmiOfBrie, Melanchthon, Mike Selinker, CoolKatt number 99999, RenamedUser jaskldjslak904, Crystallina, Burnwelk, SmackBot, Jeb R, Wcquidditch, Steam5, Chris the speller, A, Fightin-girish, New World Man, Greenshed, Grejlen, Ciller, BenH, Mattdp, Donotsayno, Rollosmokes, Mark Lungo, Whomp, BigT2006, Gatorman, Darrel M, Morgan Wick, Jcd91, Timothy Chavis, TM56, Cydebot, Northwest, Davidboyd9, الكلبه ...ايل غونج كيم انا!, Buckner 1986, Leuqarte, Chaucer1387, Acroterion, Bradhasbrouck, Realaudio2007, WCCOfan, MartinBot, SuperMarioMan, Eliz81, SeeOurZed, Coconut-Head65, Rosspz, StaticGull, ImageRemovalBot, Trivialist, 718 Bot, Mlaffs, Kjosy, Lightbot, 8VSB, Tlonca, Jesse59, MadManAmeica, Mattg82, FrescoBot, Freshh, DrilBot, Full-date unlinking bot, John123521, Mariacer Cervantes, Lazy Devil, BookDen, BornonJune8, Fairlyoddparents1234, Hotcoffee 01, Liamkasbar, DJV11181988, LR2014 and Anonymous: 68

31.6.2 Images

- **File:Ambox_important.svg** *Source:* https://upload.wikimedia.org/wikipedia/commons/b/b4/Ambox_important.svg *License:* Public domain *Contributors:* Own work, based off of Image:Ambox scales.svg *Original artist:* Dsmurat (talk · contribs)

- **File:Blank_television_set.svg** *Source:* https://upload.wikimedia.org/wikipedia/commons/8/8c/Blank_television_set.svg *License:* CC-BY-SA-3.0 *Contributors:* en:Image:Aus tv.png (among others) *Original artist:* Traced by User:Stannered

- **File:Blue_iPod_Nano.jpg** *Source:* https://upload.wikimedia.org/wikipedia/commons/c/c1/Blue_iPod_Nano.jpg *License:* Public domain *Contributors:* ? *Original artist:* ?

- **File:Cinematic-Titanic-2011-09-24-Cast.jpg** *Source:* https://upload.wikimedia.org/wikipedia/commons/0/0d/Cinematic-Titanic-2011-09-24-Cast.jpg *License:* Public domain *Contributors:* Own work *Original artist:* Evan-Amos

- **File:Cinematic-Titanic-2011-09-24-Live.jpg** *Source:* https://upload.wikimedia.org/wikipedia/en/a/aa/Cinematic-Titanic-2011-09-24-Live.jpg *License:* Fair use *Contributors:* Photo by Evan-Amos *Original artist:* ?

- **File:Clay_and_Lar'{}s_Flesh_Barn.jpg** *Source:* https://upload.wikimedia.org/wikipedia/en/f/f8/Clay_and_Lar%27s_Flesh_Barn.jpg *License:* ? *Contributors:* Screen capture from *Mystery Science Theater 3000*, "Women of the Prehistoric Planet" [0104], invention exchange segment, taken by Jeff Q *Original artist:* ?

- **File:Commons-logo.svg** *Source:* https://upload.wikimedia.org/wikipedia/en/4/4a/Commons-logo.svg *License:* CC-BY-SA-3.0 *Contributors:* ? *Original artist:* ?

- **File:CrowTRobot.JPG** *Source:* https://upload.wikimedia.org/wikipedia/commons/f/f0/CrowTRobot.JPG *License:* CC BY-SA 2.0 *Contributors:*

- MST3kBots.jpg *Original artist:*

- derivative work: Herodotus (<a href='//commons.wikimedia.org/w/index.php?title=User_talk:Herodotus,,&,,action=edit,,&,,redlink=1' class='new' title='User talk:Herodotus (page does not exist)'>talk</a>)

- **File:Darkstar_promo_flyer.jpg** *Source:* https://upload.wikimedia.org/wikipedia/en/7/7f/Darkstar_promo_flyer.jpg *License:* Fair use *Contributors:* Game's Facebook page *Original artist:* ?

- **File:Doctor_Laurence_Erhardt.jpg** *Source:* https://upload.wikimedia.org/wikipedia/en/a/ac/Doctor_Laurence_Erhardt.jpg *License:* ? *Contributors:* ? *Original artist:* ?

- **File:Edit-clear.svg** *Source:* https://upload.wikimedia.org/wikipedia/en/f/f2/Edit-clear.svg *License:* Public domain *Contributors:* The *Tango! Desktop Project.* *Original artist:* The people from the Tango! project. And according to the meta-data in the file, specifically: "Andreas Nilsson, and Jakub Steiner (although minimally)."

- **File:Felicia_Day_2012.jpg** *Source:* https://upload.wikimedia.org/wikipedia/commons/7/79/Felicia_Day_2012.jpg *License:* CC BY-SA 2.0 *Contributors:* Felicia Day *Original artist:* MingleMediaTVNetwork

- **File:Flag_of_Minnesota.svg** *Source:* https://upload.wikimedia.org/wikipedia/commons/b/b9/Flag_of_Minnesota.svg *License:* Public domain *Contributors:* ? *Original artist:* ?

- **File:Flag_of_the_United_States.svg** *Source:* https://upload.wikimedia.org/wikipedia/en/a/a4/Flag_of_the_United_States.svg *License:* PD *Contributors:* ? *Original artist:* ?

- **File:Gnome-html.png** *Source:* https://upload.wikimedia.org/wikipedia/commons/7/7e/Gnome-html.png *License:* GPL *Contributors:* www.zeusbox.org *Original artist:* zeus

- **File:GypsyMST3K.JPG** *Source:* https://upload.wikimedia.org/wikipedia/en/7/75/GypsyMST3K.JPG *License:* Fair use *Contributors:* ? *Original artist:* ?

- **File:Hadcover.jpg** *Source:* https://upload.wikimedia.org/wikipedia/en/b/b6/Hadcover.jpg *License:* Fair use *Contributors:*
  It is believed that the cover art can or could be obtained from the publisher or studio.
  Original artist: ?

- **File:Industry5.svg** *Source:* https://upload.wikimedia.org/wikipedia/commons/2/2a/Industry5.svg *License:* CC0 *Contributors:* https://openclipart.org/detail/237859/factory *Original artist:* Tsaoja

- **File:JoelRobinson.SleepyEyed.jpg** *Source:* https://upload.wikimedia.org/wikipedia/en/7/73/JoelRobinson.SleepyEyed.jpg *License:* ? *Contributors:*
  Screen capture from *Mystery Science Theater 3000*, "Jungle Goddess" [0203], invention exchange segment, taken by Jeff Q *Original artist:* ?

- **File:KLGTV23.png** *Source:* https://upload.wikimedia.org/wikipedia/en/d/d9/KLGTV23.png *License:* Fair use *Contributors:*
  The logo may be obtained from WUCW.
  Original artist: ?

- **File:Ktmalogo.png** *Source:* https://upload.wikimedia.org/wikipedia/en/e/ed/Ktmalogo.png *License:* Fair use *Contributors:* ? *Original artist:* ?

- **File:Kyle-cassidy-kevin-murphy.jpg** *Source:* https://upload.wikimedia.org/wikipedia/commons/3/38/Kyle-cassidy-kevin-murphy.jpg *License:* CC BY-SA 3.0 *Contributors:* Email *Original artist:* Kyle Cassidy

- **File:Laserblastdrforrester.jpg** *Source:* https://upload.wikimedia.org/wikipedia/en/2/2f/Laserblastdrforrester.jpg *License:* Fair use *Contributors:*
  TV
  Original artist: ?

- **File:Liner_notes_ive_seen_that_movie_too.jpg** *Source:* https://upload.wikimedia.org/wikipedia/en/3/37/Liner_notes_ive_seen_that_movie_too.jpg *License:* Fair use *Contributors:* [1] *Original artist:* Mike Ross

- **File:MST3K-Cambot.jpg** *Source:* https://upload.wikimedia.org/wikipedia/en/5/54/MST3K-Cambot.jpg *License:* Fair use *Contributors:*
  Screencap of episode
  Original artist: ?

- **File:MST3K-logo.png** *Source:* https://upload.wikimedia.org/wikipedia/en/f/fc/MST3K-logo.png *License:* ? *Contributors:*
  http://MST3K.com *Original artist:* ?

- **File:MST3KNelsonMurphy98.jpg** *Source:* https://upload.wikimedia.org/wikipedia/commons/9/9f/MST3KNelsonMurphy98.jpg *License:* GFDL *Contributors:* 1998 photograph by Infrogmation *Original artist:* Infrogmation (http://en.wikipedia.org/wiki/User:Infrogmation)

- **File:MST3K_Frank_and_Clay.png** *Source:* https://upload.wikimedia.org/wikipedia/en/9/95/MST3K_Frank_and_Clay.png *License:* Fair use *Contributors:* DVD of the episode Master Ninja II *Original artist:* Best Brains, Inc.

- **File:MST3kBots.jpg** *Source:* https://upload.wikimedia.org/wikipedia/commons/b/bf/MST3kBots.jpg *License:* CC BY-SA 2.0 *Contributors:* MST3K Robots at Super-Con *Original artist:* Steve from San Francisco, CA, USA

- **File:Merge-arrows.svg** *Source:* https://upload.wikimedia.org/wikipedia/commons/5/52/Merge-arrows.svg *License:* Public domain *Contributors:* ? *Original artist:* ?

- **File:MikeAndTheBots.JPG** *Source:* https://upload.wikimedia.org/wikipedia/en/7/78/MikeAndTheBots.JPG *License:* ? *Contributors:* ? *Original artist:* ?

- **File:Mrb5.jpg** *Source:* https://upload.wikimedia.org/wikipedia/en/c/cc/Mrb5.jpg *License:* ? *Contributors:*
  1991 episode of Mystery Science Theater 3000 showing Mr. B Natural in short during episode 319 War of the Colossal Beast
  Original artist: ?

- **File:Nuvola_mimetypes_html.png** *Source:* https://upload.wikimedia.org/wikipedia/commons/f/f9/Nuvola_mimetypes_html.png *License:* LGPL *Contributors:* http://icon-king.com *Original artist:* David Vignoni / ICON KING

- **File:Patton_Oswalt_by_Gage_Skidmore.jpg** *Source:* https://upload.wikimedia.org/wikipedia/commons/8/82/Patton_Oswalt_by_Gage_Skidmore.jpg *License:* CC BY-SA 3.0 *Contributors:* Own work *Original artist:* Gage Skidmore

- **File:Question_book-new.svg** *Source:* https://upload.wikimedia.org/wikipedia/en/9/99/Question_book-new.svg *License:* Cc-by-sa-3.0 *Contributors:*
  Created from scratch in Adobe Illustrator. Based on Image:Question book.png created by User:Equazcion *Original artist:* Tkgd2007

- **File:RAY_headshot1.jpg** *Source:* https://upload.wikimedia.org/wikipedia/commons/b/be/RAY_headshot1.jpg *License:* Public domain *Contributors:* Demorge Brown *Original artist:* Tomservotron

- **File:Rifftrax_Homepage_12_31_13.png** *Source:* https://upload.wikimedia.org/wikipedia/en/e/e0/Rifftrax_Homepage_12_31_13.png *License:* Fair use *Contributors:* http://www.rifftrax.com/ *Original artist:* Legend Films

- **File:Rifftrax_Logo.png** *Source:* https://upload.wikimedia.org/wikipedia/en/d/d7/Rifftrax_Logo.png *License:* Fair use *Contributors:* http://rifftrax.tumblr.com/ *Original artist:* ?

- **File:Rifftrax_crew_sdcc_2009.jpg** *Source:* https://upload.wikimedia.org/wikipedia/commons/7/74/Rifftrax_crew_sdcc_2009.jpg *License:* CC BY-SA 2.0 *Contributors:* The RiffTrax crew! *Original artist:* Kris Awesome from San Diego, CA, USA

31.6.3 Content license

www.ingramcontent.com/pod-product-compliance
Lightning Source LLC
Chambersburg PA
CBHW080713190526
45169CB00006B/2357